SpringerBriefs in Social Work

More information about this series at http://www.springer.com/series/13578

Sana Loue

Therapeutic Farms

Recovery from Mental Illness

 Springer

Sana Loue
Case Western Reserve University
Cleveland, Ohio, USA

ISSN 2195-3104   ISSN 2195-3112 (electronic)
SpringerBriefs in Social Work
ISBN 978-3-319-13538-0   ISBN 978-3-319-13539-7   (eBook)
DOI 10.1007/978-3-319-13539-7

Library of Congress Control Number: 2016932874

Springer Cham Heidelberg New York Dordrecht London

Printed on acid-free paper

Springer International Publishing AG Switzerland is part of Springer Science+Business Media (www.springer.com)

Preface

The original founders of the asylum movement envisioned the provision of a caring, supportive, structured environment, away from the chaotic and stressful demands of daily living, as a mechanism to help individuals recover from their mental illness. The compassionate care by asylum staff of those suffering from mental illness was often seen as a sacred duty:

> Always bear in mind that you are in your senses, and that those who are under your care are not: this is your health and happiness; that is their affliction and disease: and you cannot shew your gratitude to God for his mercy and goodness to yourself, than by shewing kindness and consideration to these your afflicted brethren. (Smith, 1999, p. 141, quoting staff instructions from the Oxford Asylum)

Such a setting, it was theorized, would allow individuals the space and time to recalibrate and regain the ability to interact with others. Indeed, the word asylum in its original usage referred to "a place offering protection and safety," or "the protection afforded by a sanctuary" (Pickett, 2000, p. 112).

The original asylum founded on these principles—"moral treatment"—gradually evolved into what is now thought of as the modern asylum, those overcrowded, bureaucratic institutions, in which individual needs succumb to staff demands for orderliness and government edicts to reduce costs. And now, when people think of an asylum, they often conjure up an image reminiscent of scenes from the book and film *One Flew Over the Cuckoo's Nest*: a decrepit, poorly ventilated building, with barred windows, locked doors, tasteless meals, and a brutal staff with little understanding of mental illness or compassion for those suffering from it (Kesey, 1962).

Therapeutic farm communities designed to aid individuals in a peaceful, supportive, caring, and therapeutic environment as they move toward recovery from mental illness continue to exist in both the United States and Europe. Today's therapeutic farms are quite diverse with respect to their underlying philosophy, their organizational and financial structure, and the services that they provide. This book is intended to acquaint its readers—mental health care and public health professionals and students, policymakers, and mental health-care consumers and their families and friends—with therapeutic farms, the services they offer, the challenges they face, and their potential to aid in recovery.

Chapter 1 focuses on the development of the therapeutic farm as a treatment modality in both the United States and in Europe, noting their different evolutionary trajectories. Chapter 2 provides an overview of the variety of services that may be offered at the therapeutic farms, which range from a structured work and vocational training program to equine therapy to yoga to medication management. Chapter 3 focuses on the organizational and financial structure of therapeutic farms. The differences noted reflect the specific legal requirements of the jurisdictions in which the farms are situated, variations in the motivations for the establishment of the farms, and the extent to which funding may or may not be available.

Chapters 4–6 provide case studies of three specific therapeutic farms for adults with mental illness: Hopewell Therapeutic Farm Community in Mesopotamia, Ohio; CooperRiis Healing Community in Mill Spring, North Carolina; and Slí Eile in Churchtown, Ireland. While all three focus on providing a setting to facilitate individuals' recovery from the effects of their mental illness, each is unique with respect to its organizational and financial structure, programming, and entrance requirements.

The final chapter examines both the strengths of therapeutic farms for adults with mental illness and the challenges that they face. Suggestions are offered for the way forward.

Unfortunately, therapeutic/care farms garner little attention and are relatively unknown until someone is actively in search of a solution to their own or their loved one's mental difficulties. Although they are not appropriate or advisable for everyone, therapeutic farms offer an alternative to hospitalization and to outpatient care. Above all, they serve as a place of respite and recovery and a source of hope for a better future.

References

Kesey, K. (1962). *One flew over thune cuckoo's nest.* New York: Penguin.
Pickett, J. (Ed.). (2000). *The American heritage dictionary of the English language* (4th ed.). Boston: Houghton Mifflin Company.
Smith, L. D. (1999). *'Cure, comfort and safe custody': Public lunatic asylums in early nineteenth-century England.* London: Leicester University.

Cleveland, OH   Sana Loue

Acknowledgments

This book would not have been possible were it not for the openness, hospitality, and sharing of numerous people involved with therapeutic farms. Foremost among them are the executive directors of three therapeutic farms: Richard Karges of Hopewell Therapeutic Farm Community in Mesopotamia, Ohio; Virgil Stucker of CooperRiis Healing Community in Mill Spring, North Carolina; and Joan Hamilton of Slí Eile in Churchtown, Ireland. The directors, staff, and residents at each of these farms met with me over periods of days and allowed me to accompany them as they went about their daily activities. They were generous with both their hospitality and their stories, helping me to develop a fuller understanding of the nature of the farms and the residents' experiences managing their mental illnesses and their recoveries. Virgil Stucker was unceasing in his efforts to ensure that I had all of the printed information that I could possibly use, in addition to providing me with ample opportunity to meet with staff and residents. Candace Carlton, the director of Quality Improvement and Compliance, and Sherry Bacon-Graves, the outcomes/program evaluation coordinator at Hopewell Therapeutic Farm Community, generously shared data (de-identified), program manuals, outcome studies, and employee manuals. I also owe much thanks to my editors at Springer, Bill Tucker and Jennifer Hadley, for their support of this project.

Contents

Chapter 1
The Development of the Therapeutic Farm

The underlying approach to and development of the therapeutic farm community has been decidedly different in large degree between the United States and Canada, on the one hand, and Western European countries on the other. Initially premised on a "moral treatment" approach to the treatment of mental illness, most, if not all, of the currently functioning therapeutic farms in the United States and Canada continue to operate from this perspective. In contrast, therapeutic farms in Western European and Scandinavian countries, such as Germany, the Netherlands, Norway, and the United Kingdom, have variously been developed specifically as an intervention for individuals with severe mental illness (or others with varying mental or physical health needs) or as an alternative mechanism for the utilization of farms, with the provision of a therapeutic experience as a secondary goal.

The European Origins of Moral Treatment

The concept of moral treatment of mental illness represented the cornerstone of care for mentally ill persons in Western Europe, the United States, and Canada during the mid-nineteenth century, with optimism for this approach reaching its zenith between 1830 and 1850 (Dain, 1964). The initial impetus for the moral treatment of mental illness derived from a belief, championed by Philippe Pinel in France and William Tuke in England, that individuals could recover from mental illness if provided with a warm and understanding environment that furnished medical treatment, occupational therapy, and religious exercises and was devoid of physical violence (Morrissey & Goldman, 1984).

Unlike their predecessors and many of their contemporaries, Pinel and Tuke viewed mental illness from a medical-psychological, rather than a theological, perspective. Pinel was keenly aware of the horrific impact of mental illness, observing that "Of all the afflictions to which human nature is subject, the loss of reason is at once the most calamitous …," often resulting in the annihilation of the individual's character and

© Springer International Publishing Switzerland 2016
S. Loue, *Therapeutic Farms*, SpringerBriefs in Social Work,
DOI 10.1007/978-3-319-13539-7_1

consciousness (Pinel, 1906, p. xv). He was critical of both "the managers of those institutions [for mentally ill persons], who [were] frequently men of little knowledge and less humanity" and of the "regular physicians [who] have indulged in a blind routine of inefficient treatment" (Pinel, 1906, p. 4). Pinel similarly rejected the then-current belief that there existed a one-to-one correspondence between the underlying cause of a mental illness and the specific nature of that illness (Pinel, 1906, pp. 14–15).

Pinel conceived of moral treatment as the use of "intimidation, without severity; of oppression, without violence; and of triumph, without outrage" (Pinel, 1906, p. 63). Kind treatment, engagement in labor, personal liberty consistent with safety, and mildness or firmness appropriate to the patient and the situation were fundamental to his approach (Pinel, 1906, p. 83). Order and moderation were deemed critical both to recovery from illness and to the management of the institution (Pinel, 1906, p. 99). He further observed:

> In the moral treatment of insanity, lunatics are not to be considered as absolutely devoid of reason, i.e. as inaccessible by motives of fear and hope and sentiments of honour …. In the first instance it is proper to gain an ascendancy over them, and afterwards to encourage them. (Pinel, 1906, p. 103)

Tuke, a Quaker, had envisioned providing care to mentally ill persons through a system of "moral management" that was to include kindness, attendance at religious services, exercise, entertainment, and occupational therapy and was devoid of then-current medical treatments such as bleeding, blisters, and evacuants (Dain & Carlson, 1960, p. 278). Treatment at the York Retreat in England, founded by Tuke in 1796, has been characterized as "one of invariable kindness, as the only rational mode of influencing the insane" (Digby, 1985, p. 33).[1] Tuke described the Retreat's reception of a new patient with a history of violence:

> Some years ago a man, about thirty-four years of age, of almost Herculean size and figure, was brought to the house … so constantly, during the present attack, had he been kept chained, that his clothes were contrived to be taken off and put on by means of strings, without removing his manacles. They were … taken off when he entered the Retreat, and he was ushered into the apartment, where the superintendents were supping. He was calm; his attention appeared to be arrested by his new situation. He was desired to join in the repast, during which he behaved with tolerable propriety … The maniac was sensible of the kindness of his treatment. He promised to restrain himself, and he so completely succeeded, that, during his stay, no coercive means were ever employed towards him … in about four months he was discharged perfectly recovered. (Tuke, 1813 [1964], pp. 146–147)

The York Retreat initially housed only three patients, although it gradually grew in size. Tuke often referred to the residents and the staff as "family" or, consistent with Quaker practice, as "friends" (Digby, 1985, p. 50; Glover, 1984, p. 43). The maintenance of a low patient-staff ratio and an intimate family environment permitted prolonged observation and individualized treatment of each resident. It was through this observation that the Retreat's physician eventually concluded that the physical treatments that had been in use actually worsened the patients' conditions (Glover, 1984, pp. 53–56).

Religion played a major role in the recovery effort. The admission of a resident population that was religiously homogeneous served to promote uniformity and wholesomeness. The Bible was read to patients on a regular basis and they were invited to attend religious services. Indeed, "it was hoped that the Religious framework that had supported Friends when sane would prove even more efficacious in meeting the greater needs of those who in some degree had lost their sanity" (Digby, 1985, p. 25). It was believed that each individual possessed an "inner light" of God that was not extinguished by mental illness and that could be nurtured (Stewart, 1992, pp. 52–53).

Tuke believed that attempts to reason with mentally ill persons would be futile due to their compromised intellectual capacity (Tuke, 1813, p. 136). Appeal to their "moral feeling" would have a greater likelihood of success. The modeling of desired behavior to patients, respectful treatment of them, and the inculcation of self-esteem were critical components of moral treatment. Self-esteem was dependent on one's esteem of others. Tuke observed:

> the patient feeling himself of some consequence, is induced to support it by the exertion of his reason, and by restraining those dispositions, which, if indulged, would lessen the respectful treatment he receives; or lower his character in the eyes of his companions and attendants. (Tuke, 1813, p. 159)

Kathleen Jones (1996, pp. vii–xv) observed that because the English term "moral treatment" is not synonymous in meaning to the French "traitement moral," the approaches of Pinel and Tuke to the treatment of mental illness cannot be said to be the same. She suggests that whereas Pinel advocated a therapy of the emotions, Tuke's conceptualization of treatment was premised on a "moral sense."[2]

The building itself was designed by a Quaker architect, John Bevans, and was to look like a family home (Edginton, 1997, p. 94). The facility was to be called "The Retreat" in order to make clear that it was used to repair the body and the mind in a safe environment (Edginton, 2003, p. 110). The grounds were

> laid out with walks, wooded glades, gardens and orchards the original eleven acres of land had been extended to twenty-seven by 1839: this formed a tranquil setting in which patients could hope to regain their serenity. (Digby, 1985, p. 43)

It has been suggested that this environment "did not place the lunatic closer to the calm influence of nature but closer to the influence of a constructed nature" that provided the patient with an opportunity to escape from the chaos of life (Edginton, 1997, p. 98). Accordingly, the asylum not only represented "a passage to sanity" but also "signified natural order (the landscape) and social order (the bourgeois community)" (Edginton, 1994, p. 377). That natural order was believed to have been divinely ordained, and the benefits to be obtained by patients from their walks in this constructed nature would provide them with both spiritual and physical benefits (Digby, 1985, p. 55).

Door locks were fashioned in such a way as to reduce noise and the doors themselves were designed to open outward in order to prevent barricading (Edginton, 1997, p. 95). Rooms were to appear cheerful, which could be accomplished through the installation of new carpeting, wallpaper, and skylights. Separate spaces were defined and maintained for the superintendent, the matron, the attendants, the servants, and the patients, reflecting "the hierarchy of a family patriarchy" (Edginton, 1997, p. 97). There were separate spaces for men and women; the separation of the sexes was maintained throughout the building and on all occasions, with the exception of religious services and concerts.

The orderliness and respect engendered through moral treatment or moral management in the asylum would ultimately facilitate patients' adaptation to society's expectations. As the physician John Conolly explained:

> Calmness will come; hope will revive; satisfaction will prevail. Some unmanageable tempers, some violent or sullen patients, there must always be; but much of the violence, much of the ill-humour, almost all the disposition to meditate mischievous or fatal revenge, or self-destruction will disappear ... cleanliness and decency will be maintained or restored; and despair itself will sometimes be found to give place to cheerfulness or secure tranquility. (Conolly, 1847, p. 143)

Moral Treatment in North America

Evolving Understandings of Mental Illness

Prior to the mid-to-late eighteenth century, during the colonial period, there had been little concern with mental illness; the focus was instead on poverty (Rothman, 1971). Protestant ethic required that the larger community assist the poor in their times of need, while simultaneously demanding that the poverty-stricken respect the social order and hierarchy that placed that placed them well toward the bottom.

Colonial communities supported their poor members in their own homes or those of their family or a neighbor; the almshouse or the workhouse was solutions of last resort (Rothman, 1971). Calvinist belief asserted that criminal acts were religious in nature and were inherent in man. Idolatry, witchcraft, and blasphemy were deemed to be not only offenses against man but against God as well, "testimony to the natural depravity of man and the power of the devil" (Rothman, 1971, p. 15). Punishment might take the form of the stocks, whipping, or death. Mental illness came to be viewed as the result of demonic possession or a form of debasement (Taubes, 1998). Mentally ill persons were treated with bleeding, cathartics, and ice baths; the less fortunate were subjected to imprisonment.

The ideas of Pinel and Tuke were brought to the United States by Benjamin Rush and Eli Todd, both of whom had attended school in England (Wood, 2004, p. 22) and were championed by Dorothea Dix (Luchins, 2001, p. 472). By 1841, a total of 16 moral treatment asylums had been established in the United States (Tomes, 1994, p. 74), modeled along the lines of those in Europe (Taubes, 1998).

The "moral treatment" of mental illness took hold in the United States in the midst of a nineteenth-century social reform movement that sought to improve the living conditions of less fortunate persons (Grob, 1966, 1973; Rothman, 1971). The practice of moral treatment, which had its heyday during the period of 1815–1875, arose from the convergence of two simultaneously occurring movements: the national Protestant revival (the "Great Awakening") that occurred between the years 1780 and 1830 and new understandings of mental illness (Taubes, 1998).

During this religious revival period, Christians were exhorted by their churches to save the souls of their community members and, indeed, to focus their efforts on saving all of humanity and purifying American society (Luchins, 1992, p. 209). The ultimate goal was nothing less than the perfection of individuals and of society through the purging of corruption and the eradication of abuses, e.g., slavery and vice, all in preparation for the coming millennium (Luchins, 1992, pp. 210–211). What was needed to prevent and remedy the unhealthful situations that could predispose individuals to mental illness was nothing less than a "healthful moral influence" (Rothman, 1971, p. 73, quoting William H. Channing, a Unitarian minister, 1844). In contrast to previous notions of the inevitability of deviance, it was now believed that man could improve his condition by engaging with greater meaning with his social and physical environment and, by doing so, transform an ill-ordered, evil life into one of stability, thereby ensuring that good and order would prevail.

As ideas of the Enlightenment took hold, understandings of crime and deviance as the result of man's nature and Satan's influence gave way to a belief that deviance—criminal acts, drinking, and other social vices—was the product of a poor family upbringing and unhealthy environment. Mental illness came to be understood as a physical illness, a disease of the brain, rather than the result of demonic possession (Taubes, 1998, p. 1002). Buttolph, a superintendent at the New Jersey State Lunatic Asylum in Trenton explained:

> The brain is composed of many regions and parts, each being endowed with the power of manifesting the several classes and individual faculties of the mind …. The mental forms of disease of the brain, of course correspond precisely with the region or part affected, and are as numerous and varied as the number and functions of such regions or parts …. As the brain in its functional office is divided into three general regions, the region of intellect, of sentiment and of animal or selfish feelings; so insanity is divided into three principal forms, which are characterized by the disturbed state of these several classes of faculties. (Buttolph, 1852, pp. 23–25, quoted in Taubes, 1998, p. 1002)

American physicians often attributed the ultimate cause of mental illness to social, emotional, and physical ailments, ranging from heart disease to a blow to the head to religious anxiety (Rothman, 1971, p. 111). Such events, it was believed, could lead to changes in the brain, thereby bringing about mental illness (Taubes, 1998). In contrast to their European counterparts, American physicians also ascribed what was seen as a growing number of cases of mental illness to individuals' pursuit of their own unrealistic attempts to achieve in an increasingly complex civilization (Rothman, 1971, p. 115).

The Adoption of Moral Treatment and the Development of the Asylum

In this context, moral treatment and the asylums built to provide care to persons deemed to be mentally ill reflected a benevolent intent, rather than an attempt to coerce and confine those who were deemed to be different (Rothman, 1971). It was believed that man could improve his condition by acting upon his social and physical environment; accordingly, mental illness could be overcome by creating and controlling the environment in such a way as to facilitate its cure.

Principles of Moral Treatment

It was believed that, despite an individual's mental illness, some part of his or her faculties remained healthy and that healthy element could be accessed through treatment. The superintendent of the State Lunatic Hospital in Worcester, Massachusetts, Samuel B. Woodward, had explained: "All the insane are in a greater or less degree, monomaniacs. But it is very rare, that all the faculties of the mind are alike affected, even in the worst form of mania …" (Woodward, 1842, p. 41, quoted in Taubes, 1998, p. 1003).

As in Europe, religion and religious services in particular were viewed as potent aids to patients' recovery efforts. The silence, structure, and calm of religious services, it was believed, could facilitate individuals' development of self-control and rational behavior, while "Christian truth" itself would provide an effective psychological intervention (Taubes, 1998, pp. 1003–1004). Thomas H. Gallaudet noted:

> So long as the insane have any exercise of their reasoning left, and any moral and religious susceptibilities to be appealed to, and no inconsiderable portion of them retain more or less of these faculties and susceptibilities, and some of them in a striking degree, so long Divine Truth with its higher motives and consolations, will be found eminently adapted to the exigencies of their unfortunate condition and one of the most salutary and efficacious means of cure. (Gallaudet, 1844, p. 29, quoted in Taubes, 1998, p. 1005)

Isolation in the sense of a respite from stress-inducing circumstances was deemed to be especially critical. As the scholar Philo noted, "[t]his process of separation—which is at once both social and spatial—has been informed by quite specific social, cultural and professional understandings of 'madness'" (Philo, 1987, p. 402). A *Report of the Commissioners of New Brunswick* (Canada) explained the basis for separating the patient from his or her family and community:

> The first and most important step is to remove the patient from his own home and all the objects he has been accustomed to see. His false notions and harassing impressions are associated in his mind with the objects exposed to his senses during the approach of his disease. His relations have become to him stale and uninteresting, and afterwards cause of angry irritation …. The most favourable situation is a retirement, where the patient will be surrounded by objects which have a composing influence. (Commissioners, 1836–1837, App. 3)[3]

Just as both Pinel and Tuke had intended to foster and sustain intimate, supportive relationships between the clients/residents and the staff, so too did the superintendents of many of the United States' first asylums. This approach necessarily required that caseloads remain small to ensure that staff had adequate time to interact with residents and provide them with needed support.

Although Pinel and Tuke had decried the use of restraints, this view was not universally accepted by all asylum administrators in the United States and Canada. John Waddell, a superintendent of the New Brunswick (Canada) asylum, argued that mechanical restraints were at times necessary for the patient's good (Francis, 1977, p. 29). Others, in both the United States and Canada, fell at opposite ends of the spectrum, with some rejecting the use of any physical restraints and others advocating for their use as a means of discipline (Francis, 1977, p. 29; Suzuki, 1995; Wood, 2004, pp. 197–202).

Programming provided within the framework of moral treatment consisted of three primary components: work, play, and worship. Separation from the social milieu in which the mental illness had first manifest was also seen as critical to recovery (Francis, 1977, p. 31). Work was believed to improve patients' sleep and physical health and would help to inculcate the moral values of independence, industry, and self-respect (Francis, 1977, p. 34). Upper class patients, however, were exempt from this requirement and, instead, were provided with recreational opportunities. Religious services were provided as an opportunity to practice restraint and appropriate public behavior (Francis, 1977, p. 32).

The Friends Asylum at Frankford, Pennsylvania, exemplified the application of moral treatment to mental illness. The facility was established in 1813 by a group of Philadelphia Quakers (Dain & Carlson, 1960, p. 278). Until 1834, admission was limited to only Quakers in an effort to maintain the uniformity advocated by Tuke. Admission was limited to a small number of individuals—30 or fewer—in order to create and maintain a family-like environment. Although patients were grouped according to their degree of mental illness and their behavior, they were not separated by socioeconomic class as was done in other asylums of the time. Admission of individuals with a proclivity toward violence was kept to a minimum (Dain & Carlson, 1960, p. 280).

Residents of the Friends Asylum were encouraged to participate in manual labor, such as farming. It was believed that such activity would serve to counteract any inclination to be lazy, would reduce the likelihood that residents would engage in morbid thoughts, and would provide them with a means of exercise. The products of these agricultural efforts would also serve as a source of food for the residents. Both men and women could also engage in gardening activities and the women could do needlework (Dain & Carlson, 1960, p. 281).

Asylum residents most frequently ate at the table with others, using knives and forks that were not chained to their tables. Although visitors were discouraged in the belief that their presence harmed rather than helped the residents, the residents were able to work and walk outdoors. Residents were rarely confined, apart from locking them in their rooms at night, since confinement would have constituted the negation of moral treatment. Because of the high degree of physical freedom, as many as 20 residents might "escape" during the course of a single year (Dain & Carlson, 1960, pp. 281–282).

Like Pinel and Tuke, the administrators of the Friends Asylum disdained the use of restraints. However, "unruly" residents could be confined to their bed, to a dark room, or in a straitjacket. "Medical" treatment, which often included seclusion, bleeding, binding, blistering, or cupping, was administered to those who were violent or disorderly (Dain & Carlson, 1960, pp. 283–284; Tomes, 1994, p. 65). "Tranquilizer chairs," which limited an individual's movement, were also utilized. By 1840, however, only cold showers were permitted as a medical means of treatment, and by 1856, opium and other drugs were used in an effort to control disruptive behavior and foster receptiveness to the receipt of moral treatment (Dain & Carlson, 1960, p. 288).

Table 1.1 Summary of Moral Treatment Principles

- Mental illness can be cured
- Patients are rational beings
- Punishment should be avoided and reward emphasized
- Physical restraint is to be avoided
- The environment must be structured, with opportunities for both labor and socialization
- Patients are to be provided with an intimate, family-like environment
- Separation/respite from the stresses of everyday life is needed to foster recovery

Treatment often required only several months or, in the case of chronically mentally ill individuals, several years (Dain & Carlson, 1960, p. 280; Luchins, 2001, p. 474; Wright, 1997, p. 145). Treatment of each resident was individualized to the extent possible. Each person's treatment regimen could be modified as he or she gradually recovered from the illness or regressed (Luchins, 2001, p. 473). Charles Evans, one of the attending physicians at the Asylum, observed:

> Even in cases where the diseased actions is [sic] the same, the manifestations of insanity are frequently so modified by the temperament, education, and social relations of individuals, as apparently to preclude the possibility of grouping them together in the same class; and demanding that the peculiarities of each be studied, and a judgment come to respecting it, based upon the first development and successive symptoms of the malady. (Dain & Carlson, 1960, p. 286, quoting Charles Evans)

The patient was deemed to have been cured when he or she no longer displayed any symptoms, was able to resist unwanted impulses, believed that he or she had been reformed, and displayed an intent to lead a productive life (Luchins, 2001, p. 473) (Table 1.1).

Moral Architecture

Moral treatment, it was believed, could not be effectuated adequately in the absence of "moral architecture" (Francis, 1977, p. 30). The building itself was considered to be an "instrument of treatment" (Scull, 1981, p. 10) and a refuge, a far cry from the confinement of mentally ill persons in whatever buildings might then be available, which often included jails, prisons, and workhouses (Edginton, 1997, pp. 92–93). The location, configuration, and appearance of the buildings, as well as the arrangement of persons and objects within those buildings, were to be considered:

> As it is found that the external appearance, as well as the internal economy of the Hospital for the Insane, exert and important moral influence … - it is a principle now generally recognized and acted on, the good taste and a regard for comfort, should characterize all the arrangements both external and internal, as calculated to induce self-respect and a disposition to self-control. (Nova Scotia Legislative Assembly, 1846, App. 32)

In both the United States and Canada, "good taste" often meant that individuals were to be housed in such a way as to prevent the distress that might be experienced by those of a higher social class at the sight and company of those of lower social strata (Francis, 1977, p. 31; Wood, 2004, pp. 145–146). In the United States, the vast majority of physicians believed that this required the segregation of mentally ill Blacks in facilities apart from the mentally ill Whites. Francis Stribling, superintendent of Virginia's Western State Hospital from 1840 to 1874, described the requirements for such a separate facility:

> The institution should be located where the climate was agreeable to the health of Blacks, and as close to where a majority of blacks in Virginia lived. It should be built on land that allowed it to be easily ventilated and kept dry. Adequate water could be obtained from an elevated source in order to pipe it inexpensively to the attic and throughout the building.
>
> Since occupational activities were required for the patients, there needed to be at least two acres of land per patient that could easily be converted into gardens and cultivated. Cheap manure should be available and also a ready market for the products produced. The hospital should be located with a view to financial advantages. It would also be desirable to have the hospital heated with a view to financial advantage. Hence, the institution should be located in one of the cities or large towns of the Commonwealth. (Stribling, 1848, p. 34)

Stribling apparently saw no contradiction between his refusal to permit the integration of Black and White clients due to "differences between them in habits, tastes, and disposition" (Wood, 2004, p. 146) and his ownership of slaves to assist with the provision of care to the facility's White patients (Wood, 2004, p. 147).

Various superintendents of state asylums, notably John Galt of the Virginia State Asylum and Merrick Bemis of the Worcester Asylum in Massachusetts, advocated for the establishment of small cottages to replace the large state institutions, a configuration that would facilitate staff-patient interaction in a family-like atmosphere (Grob, 1966). Edward Jarvis, a Massachusetts physician with a small practice treating mentally ill persons, argued that a large number of relatively small hospitals accessible to individuals in those locales were needed, rather than fewer, larger hospitals (Grob, 1978, pp. 113–114). Large hospitals, he asserted, made moral treatment almost impossible.

The Demise of Moral Treatment

The decline of moral treatment and of the asylums that were built to provide such treatment has been attributed to growing numbers of mentally ill persons, diminishing financial resources, increasingly diverse resident populations, ineffective organizational leadership, and shifting political priorities.

During the mid- to late nineteenth century, fiscal responsibility for the care of chronically mentally ill persons was shifted from local authorities to the state governments. State legislatures viewed treatment as a secondary priority; emphasis was placed, instead, on the provision of custodial care and the protection of the public at the lowest cost possible (Morrissey & Goldman, 1984, p. 787). The small, family-like moral treatment asylums were transformed into massive institutions that, for many, became a "place of permanent exile" (Slovenko & Luby, 1974, p. 225). It has been argued that the asylum became a mechanism for "the isolation of those marginal elements of the population who could not or would not conform or could not subsist in an industrial, largely laissez-faire society" (Scull, 1977, p. 348), one that was increasingly utilized by families to care for those members who had become too burdensome to bear. The American Association of Medical Superintendents, originally organized in 1845 to lessen the isolation of asylum staff and provide a source of mutual support, increasingly became autocratic (Slovenko & Luby, 1974, p. 226). Its members were accused by the medical establishment of lacking scientific knowledge. Moral treatment was soon co-opted by members of the medical profession, neurologists in particular, in an effort to create a system of care for mentally ill persons that would fall under the purview of medicine (Luchins, 2001, p. 482; Tomes, 1994, pp. 79–80; Wright, 1997, p. 138).

By the beginning of the twentieth century, many asylums had become overcrowded, with poor hygiene and inadequate resources to care for those who had been committed to their care (Fakhoury & Pribe, 2007, p. 313). Although efforts to deinstitutionalize individuals from mental health hospitals had begun in the 1930s, primarily in an effort to conserve financial resources, it was not until the 1940s and 1950s that vigorous efforts were made to decentralize mental health services and search for more community-based alternatives for mental health treatment (Goldman & Morrissey, 1985, pp. 727–728). These efforts arose from a constellation of concurrent events: exposés and reports describing the poor and often abusive conditions that existed within some mental hospitals (Deutsch, 1948; Goffman, 1961); clinical findings suggesting that long-term institutionalization led to the deterioration, rather than the improvement, of patients' functional abilities (Barton, 1966; Wing, 1962); a movement within psychiatry itself that disputed the existence of mental illness (Szasz, 1970); and increased recognition of the rights of persons with mental illness. The new medications developed to treat schizophrenia were revolutionary in their impact: improved treatment of seriously ill patients, a reduction in the length of hospital stay, and the discharge of patients from hospitals (Klerman, 1977, p. 223). The introduction into the United States of Britain's approach to mental illness— social psychiatry—provided a model for the reintroduction of mentally ill persons into the community (Jones, 1953; Klerman, 1977). The movement toward deinstitutionalization gained even greater momentum following the issuance of the Joint Commission on Mental Illness and Health report in 1961 and the congressional passage of the Community Mental Health Centers Act in 1963.

The closure of many mental health hospitals and the deinstitutionalization of long-term patients have met with both praise and condemnation. As a result of the community mental health movement, the population of state mental hospitals was reduced from approximately 560,000 in 1950 to 140,000 persons by 1980 (Goldman, Adams, & Taube, 1983). However, research indicates that some patients were discharged without adequate preparation for their transition to the community, only to be confronted with inadequate support within the community and a lack of coordination for their care (Eiklmann, Richter, & Reker, 2005; Lamb, 2001).

These noted deficiencies of the community mental health efforts led to a course correction: the provision of community support and rehabilitation (Goldman & Morrissey, 1985, p. 729; Zipple, Carling, & McDonald, 1987, p. 541). The community support and rehabilitation approach recognizes the social welfare problems that underlie mental illness, such as poverty, inability to secure employment, and homelessness. The approach emphasizes the provision of direct care to persons in the community, rather than only the prevention of chronicity (Goldman & Morrissey, 1985, p. 729).

However, the increasing prevalence of mentally ill persons in US jails, prisons, and nursing homes and among homeless persons has led some scholars to refer to the cycle of mental health care that followed as the era of homelessness and/or transinstitutionalization (Goldman & Morrissey, 1985). As of 2008, it was estimated that US prisons and jails housed 316,000 individuals with mental illness and one-half of all state and federal prisoners and 60 % of all jail inmates had mental health difficulties (Raphael & Stoll, 2013).

Therapeutic Farms in Europe and the United States Today

Europe

Like the United States, approaches to the treatment of mental illness and attribution of responsibility for the care of individuals with mental illness have evolved as understandings of mental illness have changed. Once seen as a psychological disorder that was amenable to moral treatment, mental illness was later seen during the late nineteenth and early to mid-twentieth centuries as the result of biological dysfunction (Milligan, 2000, p. 191). The unavailability of treatments during the late nineteenth and early twentieth centuries often led to the institutionalization of those with mental illness. Later, following the development and increasing availability of effective antipsychotic medications, many Western European countries experienced a movement toward deinstitutionalization, with increasing emphasis on the need for community-based support. The availability of and access to community-based services often differs greatly across localities due to variations in government priorities and spending and the extent to which voluntary organizations providing such services are able to secure ongoing funding (Milligan, 2000, p. 198).

The provision of care on today's therapeutic farms in Europe is variously known as "green care," "care farming," "farming for health," "social farming," and "green care in agriculture" (Hine, Peacock, & Pretty, 2008, p. 247). In general, care farms seek

> to provide health, social or educational benefits through farming activities for a wide range of people. These may include those with defined medical or social needs (e.g. psychiatric patients, those suffering from mild to moderate depression, people with learning disabilities, those with a drug history, disaffected youth or elderly people) as well as those suffering from the effects of work-related stress or ill-health arising from obesity. Care farming represents a partnership between farmers, health and social care providers and participants. (Hine et al., 2008, p. 247)

A distinction is drawn between care-focused care farms and farming production-focused farms. Those therapeutic communities that provide psychotherapy, counseling, and group-based treatment programs for persons with mental illness within a farm or horticultural setting are considered to be care farms due to their emphasis on the presence of health-care professions, rather than farmers (Hine et al., 2008, p. 256). Some writers have further distinguished green care farms from others, asserting that "Green Care activities do not take place in 'normal' professional farms but in institutional, therapeutic farms or gardens, and are always under the supervision of a specialized therapist" (Dessein, Bock, & de Krom, 2013, p. 53). Yet other writers have compared care farms to the village of Geel in Belgium, in which mentally ill adults are integrated into the homes of village residents and the social fabric of the village (Hassink, Eilings, Zweekhorts, van den Nieuwnehuizen, & Smit, 2010, p. 429).

Although care farms exist throughout Western Europe, Norway and the Netherlands, in particular, have well-established care farm movements. The Netherlands, for example, experienced an increase in the number of care farms from 75 in 1998 to 591 in 2005; by 2007, there were over 800 (Hassink, Zwartbol, Agricola, Elings, & Thissen, 2007, p. 21; Wilcox, 2007, p. 16). Of the 10,000 clients who received services in 2005, 1322 (approximately 13.2 %) of them received services for psychiatric issues from 221 care farms (Hassink et al., 2007, p. 28). By 2009, an estimated 39 % of all care farm clients in the Netherlands utilized the services for "mental problems," a figure that excludes those with intellectual disabilities (Dessein et al., 2013, p. 52). Approximately 500 care farms have been established in Norway (Wilcox, 2007, p. 19). The United Kingdom now has approximately 76 care farms (Hine et al., 2008, p. 255). Most farms have an association with an institution or charity.

It has been suggested that the increase in the establishment of care farms has been fueled by socioeconomic changes occurring in European agriculture and rural areas and the concomitant need to adapt to such changes (Dessein et al., 2013, p. 50). These changes include a shift from an agriculture- and manufacturing-based economy to a more service-focused economy (Woods, 2005); an increase in connectivity between rural and urban areas in terms of transportation, mobility, and knowledge (Hedberg & do Carmo, 2012; Marsden, 2007); an aging population that requires increased public spending for health (Carone & Costello, 2006); and a growing emphasis on healthy living (Lawrence & Burch, 2010). In Norway, in particular, care farms may have arisen in response to the increased emphasis in national policy on the need to provide care for mentally ill persons (Meistad & Fjeldavli, 2004, p. 11).

The majority of care farms appear to operate from one or both of two frameworks: that of public health and that of social inclusion (cf. Dessein et al., 2013, p. 53). (Various authors have differed with respect to their characterization of these frames. Chapter 3 provides additional discussion of these diverse perspectives.) The public health frame emphasizes the potential benefits that can be derived from the provision of physical and spiritual experiences in a natural setting that encompasses seasonal cycles (De Bruin et al., 2010; De Vries, 2006). The social inclusion framework recognizes that persons with mental illness, as well as others, may have been excluded from the larger society (Dessein et al., 2013, p. 53). The focus of this approach is on (1) the reintegration of excluded persons into society through activities formulated to increase their knowledge and skills, (2) the reestablishment of their ability to engage in work, and (3) the development of their self-esteem. Participation in agricultural labor is seen as a mechanism for the restoration of a structured routine and the promotion of interactive activities. The adoption of either of these frames varies across countries: the public health frame appears to predominate in Germany, Austria, and the United Kingdom, whereas the social inclusion framework is more often the basis for care farms in Ireland and Italy (cf. Dessein et al., 2013, p. 53).

A number of countries, including Belgium, the Netherlands, Norway, and Slovenia, rely primarily on a third framework: multifunctional agriculture. This approach emphasizes the cyclical rhythm of nature, the structured and caring qualities inherent within farming activities, and the tradition of providing care on farms (Dessein et al., 2013, p. 55). In this context, the vast majority of care farmers have had previous farming experience, and care farming is viewed primarily as an entrepreneurial activity (Meistad & Fjeldavli, 2004, pp. 13–14).

Care farms for adults with mental illness may provide programming in any one or more of the following domains: health care, social reintegration, education, and employment. However, the farmers themselves are counseled to refrain from engaging with clients in therapeutic interactions and, instead, are counseled to serve as role models and attachment figures (Dessein et al., 2013, p. 52; Ferwerda-van Zonneveld, Oosting, & Rommers, 2008).

Researchers conducting an interview-based study with clients, care farmers, and mental health professionals in the Netherlands reported that clients found care farms to be both less demanding than other environments because they could work at their own pace and less stigmatizing than conventional mental health services (Hassink et al., 2010). More than 40 % of the 27 participating health professionals indicated that they welcomed collaboration with care farmers because the farmers were less likely to focus on clients' limitations compared with many health-care professionals.

United States

Therapeutic farms premised on the basic principles or approach of moral treatment and focused on the recovery of adults from the symptoms of mental illness continue to exist in the United States, although they remain relatively few in number. Among those that are, perhaps, the most widely known are CooperRiis (North Carolina), Gould Farm (Massachusetts), Hopewell (Ohio), Rose Hill (Michigan), and Spring Lake Ranch Therapeutic Community (Vermont). (CooperRiis and Hopewell are the focus of Chapters 4 and 5, respectively.) The specific founding impetus, programs, processes, and extent of reliance on any therapeutic modality vary across these facilities, but the farms nevertheless share a common belief in the potential for recovery from the symptoms of mental illness and the rehabilitative effects of nature, labor, and kindness in a structured and supportive environment. Other similarities across the farms include residence on the farm by residents with mental illness together with staff and staff family members and the sharing of farm tasks by residents and staff members. And, unlike many of the European care farms that may be concerned to a greater or lesser extent with both mental illness recovery and farm productivity, the US therapeutic farms share the singular goal of helping and supporting their residents' recovery from mental illness.

The approach of Gould Farm, a therapeutic farm community situated on 555 acres in the Berkshires of Massachusetts, illustrates how the principles of moral treatment provide a framework for the farms' services, despite their differences. Like the earliest of the US therapeutic farms—but unlike some of the other currently existing therapeutic farms—the establishment of Gould Farm by William and Agnes Gould more than 100 years ago was informed by Christianity. It has not, however, limited its services to those who identify as Christian (Smith, 2014, p. xvi). Gould Farm encompasses forests, streams, and a 100-acre working farm and sufficient housing to accommodate 75 residents (Smith & Beitzel, 2014, p. 21). Gould Farm's approach emphasizes the integration of therapeutic modalities "into a life-style of kindness, community demands, traditionally motivated by non-sectarian religious impulses of service" (Smith, 2014, p. xvii). Kent D. Smith, a former executive director of Gould Farm, explained this approach:

> The mode of our life together is "working with," "playing with," "celebrating with," "eating with," "sharing with," "suffering with"—all of the things that happen when a number of people live in close proximity. We do not see ourselves as therapists, but we consider our life together "therapeutic" in the deepest sense of the word—tending toward healing and new and independent life. We do not have work-therapy, but we work; we do not have occupational-therapy, but we make things and learn skills; we do not have music therapy, but we sing and play instruments; we do not have drama-therapy, but we produce plays; we do not have recreational-therapy, but we have parties, dances, games, sports, and community celebrations. We do not have therapies as compartmentalized activities, but—by intention at least—the whole life of the community encapsulates what is meant by the term *therapy*. (Smith & Beitzel, 2014, p. xvii, quoting Kent D. Smith) (italics in original)

Despite what might seem at first reading to be a rejection of professional mental health services, Gould Farm, like the other US therapeutic farms, has licensed mental health professionals on its staff and/or contracts with licensed mental health professionals, e.g., psychologists, social workers, and psychiatrists, to provide therapeutic services to their residents (Smith & Beitzel, 2014, p. 23). Gould Farm recognizes that:

> much major psychiatric illness is probably biochemical in origin and that chemotherapy is absolutely essential to many of our guests' participation in our program.
> Above and beyond our medical understanding of schizophrenia and manic-depressive illness, we consider our guests as human beings with all of the human needs and problems and joys in which all people participate. Thus, we not only determine that guests are receiving adequate medical care including medications, we also attempt to respond to the total range of human needs in a supportive, but directive, way. (Smith & Beitzel, 2014 pp. 30–31, quoting Kent D. Smith, former executive director of Gould Farm).

Notes

1. Foucault argued that the York Retreat, rather than providing benevolent care and treatment to mentally ill persons, orchestrated their religious internment and substituted social coercion and mind control for the use of physical confinement and the use of shackles (Foucault, 1972, pp. 600–601).

2. Additional discussion relating to the vocabulary in use at the time can be found in Charland, 2007, pp. 75–78.

   Jones (1993) delineated the faulty premises from which Foucault concluded that the actions of Tuke and Pinel were not only similar—they liberated patients from their chains, they did so in the same year, and they engaged in "moral treatment"—but also arose from similar premises. Jones notes that Tuke was in England and Pinel in France and neither had had any contact with the other; their countries were at war at the time; Pinel was following political orders and Tuke acted as a matter of conscience, consistent with his religious principles; Pinel was a physician and Tuke distrusted medicine; the hospitals in which Pinel worked—Salpétriére and Bicêtre—were quite unlike Tuke's York Retreat; and, as noted, "moral treatment" is not synonymous with the French "traitement moral" (Jones, 1993, p. 172). Despite what seemed to be efforts by Tuke and Pinel to ameliorate and even improve the treatment of mentally ill persons, Foucault perceived only the continuing oppression of those who were less fortunate:

   > Freed from the chains that made it purely an observed object, madness lost, paradoxically, the essence of its liberty … It imprisoned itself in an infinitely self-referring observation. It was finally chained to the humiliation of being its own object. (Foucault, 1965, p. 265)

3. Dear and Taylor have observed:

   > The history of mental health care is a study in isolation and exclusion. Every culture appears to have had its 'madness', and to have found some method of isolating the mentally ill. These methods have sometimes been crude (as when the mad were forced to live beyond the confines of Medieval towns) and often well-intentioned (as with the development of the asylum—a true place of refuge). In all situations, however, the response has been to isolate and to exclude. (Dear & Taylor, 1982, p. 37)

   This perspective is not, however, entirely accurate. In contrast to the socio-spatial exclusion of mentally ill persons that constituted a critical component of moral treatment, the town of Geel, Belgium, has instituted and continues to maintain a system of socio-spatial inclusion of mentally ill persons, whereby they board in the homes of Geel families (Philo, 1987, p. 403).

References

Barton, R. (1966). *Institutional neurosis* (2nd ed.). Baltimore, MD: Williams and Wilkins.

Buttolph, H. (1852). *Annual reports of the officers of the New Jersey State Lunatic Asylum at Trenton, for the year MDCCCLII*. Trenton, NJ: Phillips & Boswell. Quoted in Taubes, T. (1998). "Healthy avenues of the mind": Psychological theory building and the influence of religion during the era of moral treatment. *American Journal of Psychiatry, 155*(8), 1001–1008.

Carone, G., & Costello, D. (2006). Can Europe afford to grow old? *International Monetary Fund Finance and Development Magazine, 43*(3). http://www.imf.org/external/pubs/ft/fandd/2006/09/carone.htm. Accessed 15 Feb 2015.

Charland, L. C. (2007). Benevolent theory: Moral treatment at the York Retreat. *History of Psychiatry, 18*, 61–80.

Commissioners. (1836–1837). *Report, New Brunswick. Journal of the Legislative Assembly*. Quoted in Francis, D. (1977). The development of the lunatic asylum in the maritime provinces. *Acadiensis: Journal of the History of the Atlantic Region, 6*(2), 23–38, at p. 32.

Community Mental Health Centers Act, Pub. Law 88–164 (1963).

Conolly, J. (1847). *Construction and government of lunatic asylums and hospitals for the insane*. London: John Churchill.

Dain, N. (1964). *Concepts of insanity in the United States, 1789–1865*. Brunswick, NJ: Rutgers University Press.

Dain, N., & Carlson, E. T. (1960). Milieu therapy in the nineteenth century: Patient care at the Friend's Asylum, Frankford, Pennsylvania, 1817–1861. *Journal of Nervous and Mental Disease, 131*(4), 277–290.

De Bruin, S. R., Oosting, S. J., Tobi, H., Blauw, Y. H., Schols, J. M. G. A., & de Groot, C. P. G. M. (2010). Day care at Green Care farms: A novel way to stimulate dietary intake of community-dwelling older people with dementia? *Journal of Nutrition, Health, and Aging, 14*(5), 352–357.

De Vries, S. (2006). Contributions of natural elements and areas in residential environments to human health and well-being. In J. Hassink & M. van Dijk (Eds.), *Farming for health: Greencare farming across Europe and the United States of America* (pp. 21–30). Dordrecht: Springer.

Dear, M., & Taylor, S. M. (1982). *Not on our street: Community attitudes to mental health care*. London: Pion.

Dessein, J., Bock, B. B., & de Krom, M. P. M. M. (2013). Investigating the limits of multifunctional agriculture as the dominant frame for Green Care in agriculture in Flanders and the Netherlands. *Journal of Rural Studies, 32*, 50–59.

Deutsch, A. (1948). *The shame of the states*. New York: Harcourt Brace.

Digby, A. (1985). *Madness, morality, and medicine*. Cambridge: Cambridge University Press.

Edginton, B. (1994). The well-ordered body: The quest for sanity through nineteenth-century asylum architecture. *Canadian Bulletin of Medical History, 11*(2), 375–386.

Edginton, B. (1997). Moral architecture: The influence of the York Retreat on asylum design. *Health & Place, 3*(2), 91–99.

Edginton, B. (2003). The design of moral architecture at the York Retreat. *Journal of Design History, 16*(2), 103–117.

Eiklmann, B., Richter, D., & Reker, T. (2005). For and against: A crisis of community psychiatry? *Psychiatric Prax, 32*, 269–270.

Fakhoury, W., & Pribe, S. (2007). Deinstitutionalization and reinstitutionalization: Major changes in the provision of mental healthcare. *Psychiatry, 6*(8), 313–316.

Ferwerda-van Zonneveld, R., Oosting, S. J., & Rommers, R. (2008). Green care farms for children with autistic spectrum disorder. In J. Dessein (Ed.), *Farming for health. Proceedings of the community of practice farming for health*, Ghent, Belgium, November 2007. Cited in Dessein, J., Bock, B. B., & de Krom, M. P. M. M. (2013). Investigating the limits of multifunctional agriculture as the dominant frame for Green Care in agriculture in Flanders and the Netherlands. *Journal of Rural Studies, 32*, 50–59.

Foucault, M. (1965). *Madness and civilization: A history of insanity in the age of reason*. New York: Random House, Inc.

Foucault, M. (1972). *Folie et déraison: Histoire de la folie á l'âge classique*. Paris: Gallimard.

Francis, D. (1977). The development of the lunatic asylum in the maritime provinces. *Acadiensis: Journal of the History of the Atlantic Region, 6*(2), 23–38.

Gallaudet, T. H. (1844). *The twentieth annual report of the officers of the retreat for the insane at Hartford*. Hartford, CT: Tiffany & Bunham. Quoted in Taubes, T. (1998). "Healthy avenues of the mind": Psychological theory building and the influence of religion during the era of moral treatment. *American Journal of Psychiatry, 155*(8), 1001–1008.

Glover, M. R. (1984). *The retreat at York*. New York: William Sessions.

Goffman, E. (1961). *Asylums: Essays on the social situation of mental patients and other inmates*. New York: Random House.

Goldman, H. H., Adams, N. H., & Taube, C. A. (1983). Deinstitutionalization: The data demythologized. *Hospital and Community Psychiatry, 34*, 129–134.

Goldman, H. H., & Morrissey, J. P. (1985). The alchemy of mental health policy: Homelessness and the fourth cycle of reform. *American Journal of Public Health, 75*(7), 727–731.

Grob, G. N. (1966). *The state and the mentally ill: A history of Worcester State Hospital in Massachusetts, 1830–1920.* Chapel Hill, NC: University of North Carolina Press.

Grob, G. N. (1973). *Mental institutions in America: Social policy to 1875.* New York: Free Press.

Grob, G. N. (1978). *Edward Jarvis and the medical world of nineteenth-century America.* Knoxville, TN: University of Tennessee Press.

Hassink, J., Eilings, M., Zweekhorts, M., van den Nieuwnehuizen, N., & Smit, A. (2010). Care farms in the Netherlands: Attractive empowerment-oriented and strengths-based practices in the community. *Health & Place, 16,* 423–430.

Hassink, J., Zwartbol, C., Agricola, H. J., Elings, M., & Thissen, J. T. N. M. (2007). Current status and potential of care farms in the Netherlands. *Netherlands Journal of Agricultural Science-Wageningen Journal of Life Sciences, 55*(1), 23–35.

Hedberg, C., & do Carmo, R. M. (Eds.). (2012). *Translocal ruralism: Mobility and connectivity in European rural Spaces.* Dordrecht, Netherlands: Springer Science + Business.

Hine, R., Peacock, J., & Pretty, J. (2008). Care farming in the UK: Contexts, benefits and links with therapeutic communities. *Therapeutic Communities, 29*(3), 245–260.

Joint Commission on Mental Illness and Health. (1961). *Action for mental health: Final report of the Joint Commission on Mental Illness and Health.* New York: Basic Books.

Jones, M. (1953). *The therapeutic community: A new treatment method in psychiatry.* New York: Basic Books.

Jones, K. (1993). *Asylums and after: A revised history of the mental health services: From the early 18th century to the 1990s.* London: Athlone Press.

Jones, K. (1996). Foreward. In Tuke, S. (1813). [reprinted 1994]. *Description of me retreat, an institution New-York, For insane persons of me Society of Friends. Containing an account of its origin and progress, the modes of treatment, and a statement of cases.* London: Process Press.

Klerman, G. L. (1977). Better but not well: Social and ethical issues in me deinstitutionalization of me mentally ill. *Schizophrenia Bulletin, 3*(4), 617–631.

Klerman, G. L. (1977). Mental illness, the medical model, and psychiatry. *Journal of Medicine and Philosophy, 2*(3), 220–243.

Lamb, H. R. (2001). Deinstitutionalisation at the beginning of the new millennium. *New Directions in Mental Health Services, 90,* 3–20.

Lawrence, G., & Burch, D. (2010). The "wellness" phenomenon: Implications for global agri-food systems. In G. Lawrence, K. Lyons, & T. Wallington (Eds.), *Food security, nutrition, and sustainability* (pp. 175–187). London: Earthscan.

Luchins, A. S. (1992). The cult of curability and the doctrine of perfectibility: Social context of the nineteenth-century American asylum movement. *History of Psychiatry, iii,* 203–220.

Luchins, A. S. (2001). The rise and decline of the American asylum movement in the 19th century. *Journal of Psychology, 122*(5), 471–486.

Marsden, T. (2007). Mobilities, vulnerabilities and sustainabilities. In *Key note lecture at XXII congress of the European Society for Rural Sociology, Mobilities, Vulnerabilities and Sustainabilities: New questions and challenges for Rural Europe, Wageningen, The Netherlands,* 20–24 Aug 2007. Cited in Dessein, J., Bock, B. B., & de Krom, M. P. M. M. (2013). Investigating the limits of multifunctional agriculture as the dominant frame for Green Care in agriculture in Flanders and the Netherlands. *Journal of Rural Studies, 32,* 50–59.

Meistad, T., & Fjeldavli, E. (2004). Green Care service on farms—characteristics of Norwegian enterprises. Presented at the Nordic Conference on Small Business, Tromsø, Norway, June. http://www.researchgate.net/publication/228994354_Green_care_services_on_farms_Characteristics_of_Norwegian_enterprises. Accessed 15 Feb 2015.

Milligan, C. (2000). 'Breaking out of the asylum': Developments in the geography of mental ill-health—The influence of the informal sector. *Health & Place, 6,* 189–200.

Morrissey, J. P., & Goldman, H. H. (1984). Cycles of reform in the care of the chronically mentally ill. *Hospital and Community Psychiatry, 35*(8), 785–792.

Nova Scotia Legislative Assembly. (1846). *Journals of the Legislative Assembly*. Quoted in Francis, D. (1977). The development of the lunatic asylum in the maritime provinces. *Acadiensis: Journal of the History of the Atlantic Region, 6*(2), 23–38, at p. 31.

Philo, C. (1987). "Fit localities for an asylum": The historical geography of the nineteenth-century "mad-business" in England as viewed through the pages of the Asylum Journal. *Journal of Historical Geography, 13*(4), 398–415.

Pinel, P. (1906). *A treatise on insanity: In which are contained the principles of a new and more practical nosology of maniacal disorders than has yet been offered to the public—Primary source edition. (D.D. Davis trans.)*. Sheffield: W. Todd.

Raphael, S., & Stoll, M. A. (2013). Assessing the contribution of the deinstitutionalization of the mentally ill to growth in the U.S. incarceration rate. *Journal of Legal Studies, 42*(1), 187–222.

Rothman, D. J. (1971). *The discovery of the asylum: Social order and disorder in the new republic*. Boston, MA: Little, Brown and Company.

Scull, A. T. (1977). Madness and segregative control: The rise of the insane asylum. *Social Problems, 24*(3), 337–351.

Scull, A. T. (Ed.). (1981). *Madhouses, mad-doctors and madmen: The social history of psychiatry in the Victorian era*. Philadelphia, PA: University of Pennsylvania Press.

Slovenko, R., & Luby, E. D. (1974). From moral treatment to railroading out of the mental hospital. *Bulletin of the American Academy of Psychiatry and Law, 2*(4), 223–236.

Smith, S. K. (2014). Introduction. In S. K. Smith & T. Beitzel (Eds.), *One hundred years of service through community: A Gould Farm reader* (pp. xv–xx). Lanham, MD: University Press of America, Inc.

Smith, S. K., & Beitzel, T. (Eds.). (2014). *One hundred years of service through community: A Gould Farm reader*. Lanham, MD: University Press of America, Inc.

Stewart, K. A. (1992). *The York Retreat in light of the Quaker way*. York, UK: Erbor Press, William Sessions Ltd.

Stribling, F. T. (1848). Annual report. Quoted in Wood, A. D. (2004). *Dr. Francis T. Stribling and moral medicine: Curing the insane at Virginia's Western State Hospital: 1836–1874*. GallileoGianinny Publishing.

Suzuki, A. (1995). The politics and ideology of non-restraint: The case of the Hanwell asylum. *Medical History, 39*, 1–17.

Szasz, T. (1970). *The manufacture of madness*. New York: Harper and Row.

Taubes, T. (1998). "Healthy avenues of the mind": Psychological theory building and the influence of religion during the era of moral treatment. *American Journal of Psychiatry, 155*(8), 1001–1008.

Tomes, N. (1994). *The art of asylum-keeping: Thomas Story Kirkbride and the origins of American psychiatry*. Philadelphia, PA: University of Pennsylvania Press.

Tuke, S. (1813). *Description of the retreat, an institution near York, for insane persons of the Society of Friends. Containing an account of its origin and progress, the modes of treatment, and a statement of cases*. London: Process Press [1996].

Wilcox, D. (2007). Farming and care across Europe: A Nuffield Scholarships Trust Award. http://www.carefarminguk.org/sites/carefarminguk.org/files/Nuffield_Report.pdf. Accessed 30 Jan 2015.

Wing, J. (1962). Institutionalism in mental hospitals. *Journal of Social and Clinical Psychology, 1*, 38–51.

Wood, A. D. (2004). *Dr. Francis T. Stribling and moral medicine: Curing the insane at Virginia's Western State Hospital: 1836–1874*. GallileoGianinny Publishing.

Woods, M. (2005). *Rural geography*, London: Sage Publications.

Woodward, S. B. (1842). *Ninth annual report of the trustees of the State Lunatic Hospital at Worcester*. Boston, MA: Dutton & Wentworth. Quoted in Taubes, T. (1998). "Healthy avenues of the mind": Psychological theory building and the influence of religion during the era of moral treatment. *American Journal of Psychiatry, 155*(8), 1001–1008.

Wright, D. (1997). Getting out of the asylum: Understanding the confinement of the insane in the nineteenth century. *Social History of Medicine, 10*(1), 137–155.

Zipple, A. M., Carling, P. J., & McDonald, J. (1987). A rehabilitation response to the call for asylum. *Schizophrenia Bulletin, 13*(4), 539–544.

Chapter 2
Programmatic Approaches

A wide range of therapeutic programming is available at the therapeutic farms of the United States and the care farms of Western Europe and Scandinavia. The specific offerings differ across the various farms; availability is often dependent upon the needs of the facility's consumers; the ability of the facility to identify and retain a professional skilled in the use of a particular modality, either as an employee or as a consultant; the familiarity of a facility's staff with a specific approach to treatment; and the availability of funding to cover the cost of a specific service. As an example, some therapeutic farms offer services only in conjunction with residential care, some offer only day services, and still others provide both residential and day programming.

This chapter focuses on the diverse approaches and services that are utilized by care farms/therapeutic farms in the United States, Canada, and Europe for their residents. Although every attempt has been made to identify the various modalities employed, it is likely that some that are described in this chapter are no longer offered at the time of this book's printing and others that have not been included are in use. A brief discussion of the underlying theory and empirical research supporting each of these modalities is provided. The scope of this work precludes an in-depth examination of each approach.

Cognitive Behavioral Therapies

According to the National Association of Cognitive-Behavioral Therapists (2014), cognitive behavioral therapy (CBT) is an umbrella term that refers to multiple approaches characterized by their emphasis on the role of thinking as an important factor in individuals' feelings and actions. These approaches include dialectic behavior therapy (DBT), emotive behavior therapy, rational behavior therapy, rational living therapy, and cognitive therapy. The scope of this chapter does not permit

© Springer International Publishing Switzerland 2016
S. Loue, *Therapeutic Farms*, SpringerBriefs in Social Work,
DOI 10.1007/978-3-319-13539-7_2

a review or detailed description of each of these diverse approaches and the extensive body of literature that relates to each.

In general, CBT approaches focus on helping the individual identify negative or false beliefs and testing or restructuring them so that they are more aligned with the actual facts of a situation. Dialectic behavior therapy emphasizes the need to accept thoughts and feelings that may be uncomfortable and to find a balance between acceptance and change. In comparison with other therapeutic approaches, CBT tends to be briefer, involving only 16–20 individual sessions (National Association of Cognitive-Behavioral Therapists, 2014; Rector & Beck, 2001). CBT sessions are structured and directive and clients are assigned reading and homework exercises following their sessions.

One example of the type of exercise that might be used during a CBT session or as homework focuses on helping the individual identify a triggering event, the negative beliefs or thoughts that he or she automatically associates with that event, and the negative feelings that follow. The individual then examines the negative beliefs or thoughts to determine their level of accuracy and, by replacing the automatic negative thought or belief with a more positive one, is better able to regulate his or her emotional responses. As the following example demonstrates, CBT is premised on the idea that it is not the situation that causes the individual to experience negative emotions but rather the individual's interpretation of that antecedent event. Table 2.1 provides an example of how this technique might be used.

The automatic negative thoughts often derive from individuals' core beliefs about themselves, e.g., "I have nothing to offer anyone," "Nobody loves me," and "I have no friends." CBT exercises that might be utilized include keeping a worksheet that utilizes the format of Table 2.1 to track events, negative thoughts, emotions, evidence, and positive replacement thoughts and developing a "ladder" of goals that are specific, measurable, achievable, relevant, and time limited. An exercise that is often used to improve individuals' problem-solving skills involves a sequence of steps: identifying the problem and all possible solutions, evaluating negative and positive aspects of the possible solutions, selecting a solution, planning the implementation of the solution, implementing the solution, and reviewing the process and the outcome (Hertfordshire Partnership, NHS Foundation Trust n.d.).

Table 2.1 Example of cognitive behavioral exercise

Event	Automatic negative thought	Emotion	Evidence that contradicts negative thought	Positive thought replacement
Colleague passes me in the hall and does not respond when I say "Good morning"	Colleague doesn't like me; I must have offended him somehow; I have no friends here	Hurt, anger, shame	Other people have acknowledged my greeting; I go out to lunch with coworkers on some days	Maybe my colleague didn't hear me or doesn't feel well or was preoccupied with an issue and for that reason did not respond

Cognitive behavior therapy has been used effectively with individuals diagnosed with schizophrenia and with a dual diagnosis of schizophrenia and substance use disorder (Barrowclough et al., 2001; Rector & Beck, 2001), bipolar disorder (Scott et al., 2006), and moderate to severe depression (DeRubeis et al., 2005). Dialectical behavior therapy has been found to be effective in reducing symptoms of depression (Lynch, Morse, Mendelson, & Robins, 2003). However, the degree to which cognitive behavioral therapy is effective has been found to depend on the level of the therapist's expertise (DeRubeis et al., 2005).

Horticultural Activities, Therapeutic Horticulture, and Horticultural Therapy

Therapeutic farms may offer any one or more of various horticultural-related services: horticultural activities, therapeutic horticulture, and/or horticulture therapy. Horticulture activities serve as a form of recreation. In contrast, therapeutic horticulture is "a process that uses plant-related activities through which participants strive to improve their well-being through active or passive involvement" (GrowthPoint, 1999, p. 4). These activities can require active engagement, such as sowing, potting, and weeding, or may be more passive, such as watching birds and butterflies. Horticultural therapy is "the use of plants as a therapeutic medium by a trained professional to achieve a clinically defined goal" (Kam & Siu, 2010, p. 80).

These various activities occur in different settings. Horticultural activity, for example, can occur in one's own garden or in community gardens. The health benefits, which are largely incidental to the activity, may include an improved sense of overall well-being and improved physical health. Horticulture activities can also take place in the context of work, such as work on a green farm. Horticultural therapy, however, occurs in healing gardens, care institutes, or practice gardens of horticultural therapists (Elings, 2006, p. 45).

Horticulture activity itself has been found to be associated with improved coping (Unruh, 2004) and may provide participants with a sense of accomplishment and productivity (Wiesinger, Neuhauser, & Putz, 2006) Additionally, as described in greater detail on the section relating to physical activity (this chapter), gardening may have positive effects on physical and mental health (Fjeld, Veiersted, & Sandvik, 2002). Individuals involved in therapeutic horticulture may benefit by learning new skills that can then be transferred to employment settings (Sempik & Aldridge, 2006).

Research findings suggest that horticultural therapy may be associated with reductions in symptoms of depression, anxiety, and stress. One study involving 18 men and women ranging in age from 27 to 65 years, all of whom had diagnoses of major depression or dysthymia or were in the depressive phase of bipolar II disorder, found that depressive symptoms decreased during a 12-week program of horticulture therapy (Gonzalez, Hartig, Patil, & Martinsen, 2011).

Various theories have been proposed to explain the beneficial effects of horticulture therapy. Kaplan (1992) hypothesized that nature provides a restorative experience by providing the individual with a sense of escape from the ordinary part of life (being away), attracting individuals' attention involuntarily (fascinates), giving individuals a sense that they are in a well-ordered world characterized by meaning (extent), and fitting the activity of the individual (compatibility). Ulrich and Parsons (1992) theorized that individuals are so overloaded with the stimuli that confront them each day—noise, color, movement—that they become overwhelmed and over-stimulated. Because plant environments are less complex, they may help to alleviate feelings of stress. Other theories proposed by Ulrich and Parsons (1992) include one that is premised on learning theory—that individuals' responses to plants derive from their early learning experiences or the culture in which they were raised—and on evolution theory—that individuals' psychological responses to plants mirror our evolution in environments that consisted primarily of plant life.

It appears that many, if not most, of Europe's care farms and the United States' therapeutic farms offer some form of horticulture activity and/or therapeutic horticulture. Relatively few offer horticultural therapy.

Animal-Assisted Activity, Intervention, and Therapy

Care farms and therapeutic farms provide numerous and varied opportunities for their residents to interact with animals. Contact may occur in the context of providing care and tending to farm animals, such as cows, horses, and chickens, as part of one's daily routine and chores, and/or through interaction with one or more animals that are considered to be members of the household, such as a cat or dog. In the latter situation, the animal may be thought of as a companion animal, i.e., one that "is frequently in the company of, associates with, or accompanies another or others; one that assists with and lives with another as a helpful friend" (Jorgenson, 1997, p. 250).

The inclusion of animals in the daily living and daily routines in therapeutic and care farms—animal assisted intervention—should not be confused with animal-assisted therapy, animal-assisted activity, or reliance on service animals. Although these latter services may also be provided on therapeutic and care farms, it appears that they are offered less frequently. An animal-assisted intervention is a "therapeutic intervention that intentionally includes or incorporates animals as part of the therapeutic process or milieu" (Kruger, Trachtenberg, & Serpell, 2004, p. 4). In comparison, "[a]nimal-assisted therapy involves the carefully planned and monitored use of the therapist's companion animal in sessions to build rapport, enhance the therapeutic process, and facilitate positive change" (Walsh, 2009, p. 466). Animal-assisted activities are provided by trained professionals, paraprofessionals, or volunteers; neither treatment goals nor detailed clinical notes are maintained (Animal Assisted Intervention International, 2013; Kruger and Serpell, 2006). Service animals are trained to live and work in partnership with their humans in order to improve the human's quality of life and ability to function (Walsh, 2009, p. 466).

The findings from various studies suggest that the development of an animal-human bond may have direct positive effects on individuals' physical and mental health. Positive physical benefits include decreased blood pressure and heart rate (Baun, Bergstrom, Langston, & Thomas, 1983), increased survival (Friedmann, Katcher, Lynch, & Thomas, 1980), and a reduction in minor health problems (Serpell, 1990).

It has been hypothesized that animals fulfill humans' most basic needs of companionship and safety (Levinson, 1982) and provide nurturance and a consistent bond during periods of emotional upheaval and life transitions (Zasloff & Kidd, 1994). Animals may be significant attachment figures in individuals' lives (Rynearson, 1978) and may represent their only uncomplicated relationship (Quindlen, 2007). Animals may serve as confidants to hear individuals' most closely held feelings (Mallon, 1994, pp. 466–467). Indeed, an animal may serve as a "lifeline" for individuals who are particularly vulnerable (Levinson, 1970), even reducing the likelihood that an individual will attempt suicide (Hafen, Rush, Reisbig, McDaniel, & White, 2007). As an example, in discussing the bond that forms between individuals and their pets, Friedmann and colleagues explained:

> Exchanges of affection or attention between persons and their pets can take place with or without words by these persons. The speechless kind of companionship shared with pets may provide a source of relaxation that human companions who demand talk as the price of companionship may not provide. (Friedmann et al., 1980, p. 310)

A study examining the relationship between children placed in a residential treatment center and their interactions with farm animals revealed that contact with the farm animals helped to relieve children's sadness, ameliorate feelings of anger, facilitate increased calm, and develop nurturing behaviors (Mallon, 1994). Animals have been shown to alleviate depression in individuals with AIDS (Siegel, Angulo, Detels, Wech, & Mullen, 1999) and dementia (Colombo, Buono, Smania, Raviola, & De Leo, 2006), while contact with birds has been found to reduce depression, loneliness, and low morale among elderly persons in skilled rehabilitation units (Jessen, Cardiello, & Baun, 1996). Even gazing at fish in a fish tank has been found to provide greater calming benefits to individuals with dementia than does meditation (Filan & Llewelltn-Jones, 2006).

Interaction with horses, in particular, may instill individuals with a sense of confidence and a feeling of freedom (Mallon, 1994, pp. 463–464) and challenge an individual to develop a sense of courage (Hassink, 2002, p. 336). Therapeutic riding, which "uses equine-assisted activities for the purpose of contributing positively to cognitive, physical, emotional and social well-being of people with disabilities" (Professional Association of Therapeutic Horsemanship International, 2015), has been found to be beneficial to individuals who have been diagnosed with mental illness (Bizub, Joy, & Davidson, 2003). Equine-facilitated psychotherapy, which is facilitated by a credentialed, trained mental health professional in collaboration with a credentialed equine professional, may include such activities as handling, grooming, riding, and vaulting, among others (Professional Association of Therapeutic Horsemanship International, 2013).

Among individuals with schizophrenia, animals may help to decrease their apathy, increase their motivation, and improve their quality of life (Barker & Dawson, 1998; Beck, 2005). A randomized controlled trial involving 90 individuals with schizophrenia and other serious mental illnesses found that compared to those who received only standard psychiatric care, individuals who worked with farm animals for a period of 12 weeks in addition to receiving standard psychiatric care experienced improvements in coping, confidence, and quality of life (Berget, Ekeberg, & Brastad, 2008). Activities with the animals included physical contact with them and moving, cleaning, feeding, and/or milking the animals.

Some research findings suggest that a beneficial effect from interaction with the farm animals is possible only if the resident is able to form a bond with the animal(s) (Hassink, 2002, p. 336). That bond can potentially be formed with dogs, cats, horses, cows, chickens, sheep, goats, and pigs.

The farmers of some European care farms have made special efforts to ensure that their residents are able to engage in activities with the farm animals while still maintaining an adequate level of production:

> Other farmers take up the challenge to really integrate social care and agricultural production in such a way that clients can participate in the production part of the farm, without frustrating the efficiency of agricultural production. This requires a lot of courage and adaptations of the production system. Some farmers change from cows to goats because they are easier to handle for the clients, or they change the stables in such a way that the clients can feed the cows, pigs or chickens. (Hassink, 2002, p. 334)

Engagement with the farm animals helps to promote a sense of responsibility, encourages physical activity, and reduces the residents' focus on their own problems (Hassink, 2002, pp. 336–337).

Resident-animal contact is not without risk to either the animal or the resident. Individuals who have poor impulse control or who are aggressive may relieve their frustrations or anger by abusing the animal(s) (cf. Felthous & Kellert, 1987; Rigdon & Tapia, 1977; Tapia, 1971). The risk of transmission of infections from animals to the farm residents—zoonoses—may increase with increased animal contact (Hassink, 2002, p. 339). Prevention of transmission of infections such as ring scabies of cows or infection with parasites requires that residents receive appropriate instruction regarding hygiene.

Expressive Therapies: Art, Music, Movement, Dance, and Sandplay

Art Activities, Art as Therapy, and Art Therapy

The broad concept of art for health encompasses a wide range of art activities that may yield a health benefit (Daykin, Byrne, Soteriou, & O'Connor, 2008). Participation in art activities allows individuals to experience and express emotions that they might not otherwise be able to access; express their individuality; acknowledge and address anxieties (Ellingson, 1991); and overcome feelings of despair, isolation, and alienation (Kramer, 2000, p. 59). Participation in art activities by individuals with mental health needs has been found to help them develop a greater sense of empowerment (Hacking, Secker, Spandler, Kent, & Shenton, 2008); communicate thoughts, feelings, and experiences (Morter, 1997; Skailes, 1997); engage in decision-making; develop an improved sense of self-worth (Skailes, 1997); exercise self-discipline; and overcome a sense of emptiness (Molloy, 1997).

In contrast to art activities, art therapy is a process through

> which clients, facilitated by the art therapist, use art media, the creative process, and the resulting artwork to explore their feelings, reconcile emotional conflicts, foster self-awareness, manage behavior and addictions, develop social skills, improve reality orientation, reduce anxiety, and increase self-esteem. A goal in art therapy is to improve or restore a client's functioning and his or her sense of personal well-being. (American Art Therapy Association, 2013)

The process of making the art helps the client develop "a psychic organization that is sufficiently resilient to function under pressure without breakdown or the need to resort to stultifying defensive measures" (Kramer, 2000, p. 18). Importantly, the use of art in therapy, whether as the primary therapeutic modality or as an adjunctive therapy, helps the client bridge his or her inner and outer realities to produce a new entity (Ulman, 1975, p. 13).

Art therapy has been used extensively with individuals with a history of psychosis (Wood, 1997). The approach utilized in art therapy with clients with psychotic symptoms appears to mirror the principles that underlie moral treatment: respect and caring, with an emphasis on the possibility of recovery. As the art therapist E.M. Lyddiatt explained:

> Diagnosis and other medical procedures are not the concern of an art department. A person may be suffering from a disease that is labeled schizophrenia, but he remains a human being—his mind still works as do other minds, although certain notions may have become exaggerated. In essence he is unaltered and frequently can be restored to health. (Lyddiatt, 1972, pp. 5–6)

(It should be noted that some art therapists do, however, rely on the art products of their clients as a basis for diagnosis. See Lloyd, Wong, and Petchkovsky, 2007; Miller, 1998.)

It has been suggested that individuals will be more likely to participate in art therapy and express their inner reality if they have access to "a planned therapeutic environment where art materials are freely available and an art therapist is providing encouragement, support or sympathetic non-interference" (Edwards, 1978, p. 14). Lloyd et al. (2007) found from their interview study with eight adults with serious mental illness that participation in art therapy facilitated individuals' ability to access their emotions, helped them to discover purpose and meaning, provided a sense of self-discovery, increased their sense of control, aided in the development of a stronger sense of identity, and provided a source of self-satisfaction.

Music and Music Therapy

As with art, therapeutic farm programs may utilize music as a recreational activity and/or in the context of music therapy. Research findings indicate that music may help individuals identify their own emotions and synthesize, control, and modulate their emotional behavior (Thaut, 1990). Music can also be used to create an environment that facilitates the disclosure of feelings, thoughts, and problems (Harper & Bruce-Sanford, 1989). Because music stimuli can affect individuals' state of readiness to perceive, individuals may more willingly search for new perceptual experiences in the presence of music stimuli. This can occur because the behavior of perceptual curiosity is connected to and driven by the experience of reward through stimulus reduction or arousal increment (Berlyne & Borsa, 1968). Music may also reduce individuals' attention to external stimuli and facilitate increased attention to the information contained in the music (Hernandez-Peon, 1961; Marteniuk, 1976); this may be particularly important in the context of providing educational information, e.g., about mental illness. Individuals may also develop positive emotional associations when such learning is accompanied with music (Bunt, 1988).

Music therapy has been defined as "the use of music in clinical, educational and social situations to treat clients or patients with medical, educational, social or psychological needs" (Wigram, Pedersen, & Bonde, 2002, p. 29, quoting Wigram, 2000). Music therapy can be utilized for nontherapeutic but associated purposes, in conjunction with other treatment modalities, or as the primary modality of therapy (Bruscia, 1998). When used at an intensive level as the primary mode of therapy, music therapy may help individuals develop or improve their interpersonal skills and their ability to connect and communicate with others, feel empathy, and/or interact socially (Bruscia, 1998) and may bring about behavior change and transformation.

Research findings indicate that music may be helpful as an intervention with individuals experiencing symptoms of severe mental illness. Music therapy may help to reduce the negative symptoms of schizophrenia and the overall psychotic symptoms, improve the individual's social functioning, decrease his or her social isolation, and increase his or her interest in external events (Gold, Heldal, Dahle, & Wigram, 2005; Hayashi et al., 2002; Talwar et al., 2006; Tang, Yao, & Zheng, 1994). Loue, Mendez, and Sajatovic (2008) found in their study involving 53 women between the ages of 18 and 50 with diagnoses of schizophrenia, bipolar disorder, or major depression that almost one-half of the women listened to music regularly. Some of the women reported that music was essential to their lives and improved their mental and social well-being by facilitating expression and reflection of their emotions and increasing their energy levels. These findings suggest that music may affect the core negative symptoms and compensate for neuropsychological deficits in women with schizophrenia and related conditions by facilitating the articulation of emotion and allowing individuals to better attend to and potentially incorporate external activities into their lives.

Care farms and therapeutic farms use music in a variety of ways. At some of these farms, residents who are able to play musical instruments may do so for their own relaxation and/or for others' entertainment. As some farms, such as Hopewell (see Chap. 4), residents have formed a Hopewell band and practice together on a regular basis. Therapists may use music to help guide meditation sessions, combining the therapeutic effects of both music and meditation. Music lyrics may also be utilized to help individuals recognize and express their feelings or mediate their moods.

Sandplay Therapy

To the best of this author's knowledge, sandplay therapy is rarely included in therapeutic farms' programming. One notable exception is that of Hopewell, which is currently conducting research to assess the acceptability of this therapeutic modality to its residents and the extent to which it confers therapeutic benefit.

Sandplay therapy was developed by Dora Kalff (1904–1990), a Swiss psychotherapist, through her integration of concepts derived from Margaret Lowenfeld's "Worldtechnik," Jungian theory, and Buddhist contemplative practices. Lowenfeld had provided her child clients with miniature toys, sand, and water, which they could use to create what she termed "sand worlds" (Hutton, 2004). These sand worlds provided the child with a protected space to which a child could return and in which a child could freely communicate through the use of his or her senses and bodily action and play. This play, Lowenfeld believed, reflected the child's inner reality, which the therapist was to receive unconditionally (De Domenico, 1999). Jung (1928) believed that the ego represents the center of consciousness and the self represents psychological integrity or wholeness; this includes both the conscious and the unconscious.

Kalff utilized Lowenfeld's "Worldtechnik" and Carl Jung's individuation process to create the therapeutic modality that she labeled "sandplay." Like Lowenfeld, Kalff conceived of the sandtray as a "free and protected space." In contrast to Jung, who had used clients' dreams in their analytical process, Kalff utilized clients' series of sandtrays through which the clients' unconscious matter could become better known (Amman, 1991, p. xv).

Sandplay therapy offers the potential for significant benefits, including the facilitation of healing, the promotion of the individuation process, the creation of a bridge between the client's conscious and unconscious worlds, the diminution of defenses through the provision of a free and protected space, and the opportunity to engage in spontaneous, creative play (Boik & Goodwin, 2000).

Sandplay therapy utilizes a rectangular sandplay, whose bottom and sides are painted blue. The blue color provides the semblances of water or sky. The therapist maintains numerous miniature figures and small objects on open shelves and/or in drawers; the client is free to select from among these for use in the sandtray. These objects and figures are symbols that the client can use to create a world in the sand (Weinrib, 1983, pp. 11–12). The miniatures figures are often arranged by category so that the client can more easily find what he or she needs: buildings; trees, bushes, and other plants; fences, gates, and doors; transportation; wild animals; domestic animals; people; fantasy animals; fantasy and folkloric people; and multipurpose materials, such as string and paper (Thomson, 1981). The sandtray itself has been compared to a "soul garden" and an "in-between space" that promotes the development and revelation of the client's inner and outer lives (Amman, 1994).

The sandplay process frequently proceeds in silence as the client molds the sand and/or selects the objects that he or she will place in the tray. Silence can be important therapeutically because it facilitates the release of unconscious psychological material and creates a communion between the therapist and the client (Rajski, 2003). After the client has finished making the tray, the therapist may ask the client to tell a story about the tray or may ask the client, "How do you feel looking at the tray?" However, the therapist does not interpret the tray for or to the client at this time or confront the client but instead will evaluate "the picture in light of Jungian symbology and any archetypal amplifications that suggest themselves" (Weinrib, 1983, p. 13). The therapist's evaluation is only a hypothesis, because it is not shared with the client and, as a result, cannot immediately be affirmed, rejected, or modified.

The therapist photographs the sandtray after the client leaves and keeps these photographs with his or her clinical notes. Because the therapist does not dismantle the sandtray while the client is still there, the client is able to retain the image of what he or she did during the session. The therapist does not share the photographs with the client during the course of the client's therapy but may review them with the client following the termination of the therapy. This review process may help to reinforce any changes and enhance the client's understandings.

Physical Activities

Depending upon the resources available at a particular therapeutic farm, physical activity can take any one or more of several forms. To some extent, physical activity is necessary if residents are to participate in the farm activities, such as gardening, cleaning animal stalls and chicken coops, and feeding the animals. The setting of the farms may also provide opportunities for physical exercise across the four seasons: walking on trails, cross-country skiing, snowshoeing, sledding, etc. Participation in any form of physical activity may be especially important for individuals with serious mental illness due to the increased risk of ischemic heart disease, hypertension, diabetes, and respiratory disease associated with mental illness and/or various medication regimens for its treatment (Bartsch, Shern, Feinberg, Fuller, & Willett, 1990; Berren, Hill, Merikle, Gonzales, & Santiago, 1994; Koran et al., 1989; Lawrence, Holman, Jablensky, & Hobbs, 2003; Lean & Pajonk, 2003; Piette, Richardson, & Valenstein, 2004; Sernyak, Leslie, Alarcon, Losonczy, & Rosenheck, 2002).

Participation in physical activities may have a beneficial effect on both physical and mental health. One study conducted in Germany involving 7124 noninstitutionalized persons between the ages of 18 and 79 years found that within the previous year 15.8 % of the participants had substance abuse or dependence disorders, 11.9 % had affective disorders, and 14.5 % had anxiety disorders (Schmitz, Kruse, &

Kugler, 2004). Among the individuals with any of these disorders, those with lower levels of physical activity also reported a lower health-related quality of life, e.g., cardiovascular conditions, chronic respiratory diseases, and dermatologic conditions.

Numerous cross-sectional and prospective studies have found that men and women who are physically active experience decreased levels of depression and anxiety symptomatology (Allgower, Wardle, & Steptoe, 2001; Bhui & Fletcher, 2000; Camacho, Roberts, Lazarus, Kaplan, & Cohen, 1991; Dimeo, Bauer, Varahram, Proest, & Halter, 2001; Dunn, Trivedi, & O'Neal, 2001; Framer et al., 1988; Martinsen, 1994; Martinsen, Hofart, & Solberg, 1989; Rehor, Dunnagan, Stewart, & Cooley, 2001; Stephens, 1988; Strawbridge, Deleger, Roberts, & Kaplan, 2002; Van Gool et al., 2003; Wassertheil-Smoller et al., 2004; Wyshak, 2001). One study involving 5451 men and 1277 women found that there was an inverse relationship between their level of usual physical activity and their well-being. Specifically, increases in individuals' levels of cardiorespiratory fitness and usual level of physical activity were found to be associated with fewer symptoms of depression and increased well-being (Galper, Trivedi, Barlow, Dunn, & Kampert, 2006). The National Comorbidity Study, conducted in the United States between 1990 and 1992 with a national probability sample of 5877 noninstitutionalized adults ages 15–54 years, found that individuals who participated in regular physical activity had a reduced likelihood of having major depression, agoraphobia, panic attacks, generalized anxiety disorder, specific phobia, and social phobia (Goodwin, 2003).

Research findings also suggest that exercise may be an effective treatment for major depressive disorder of mild to moderate severity, for schizophrenia, and for anxiety disorders. In a study conducted between 1998 and 2001, participants ages 20–45 years with diagnoses of mild to moderate major depressive disorder were randomized for a period of 12 weeks to a placebo group or to one of four aerobic exercise treatment groups that differed by total energy expenditure and frequency (Dunn, Trivedi, Kampert, Clark, & Chambliss, 2005). The investigators reported that compared to individuals randomized to exercise less and those who received placebo, individuals who had been randomized to participate in exercise at the level and frequency recommended by public health guidelines experienced a greater decline in depressive symptoms as measured by their scores on the 17-item Hamilton Rating Scale for Depression. Yet another investigation that reviewed studies involving exercise interventions for individuals with schizophrenia concluded that exercise could alleviate symptoms of depression, low self-esteem, and social withdrawal and could assist individuals to cope with symptoms such as auditory hallucinations (Faulkner & Biddle, 1999; Faulkner & Sparkes, 1999). And in a randomized trial involving 79 male and female adults with various anxiety disorders, researchers found that both participants assigned to the 8-week walking/jogging group and those assigned to the strength/flexibility group experienced reductions in their anxiety scores (Martinsen et al., 1989).

Exercise/physical activity may confer important mental health benefits even at low levels of intensity and frequency. Researchers conducting the Scottish Health Survey, involving 19,842 men and women, observed mental health benefits even among participants who exercised only 20 min each week (Hamer, Stamatakis, & Steptoe, 2009). Any form of daily physical activity, including housework, gardening, walking, and sports, was found to be associated with a lower risk of psychological distress, even after controlling in the analyses for age, sex, socioeconomic status, marital status, body mass index, chronic illness, smoking, and year of survey.

In addition to activities that utilize the farm surroundings, such as nature walks, formal programming may include physical activities that are provided in a group setting but do not involve group interaction, such as aerobics or yoga. Physical activity can also be structured to address some of the psychosocial consequences that often accompany mental illness such as social isolation. It has been suggested, for example, that participation in physical activity offers a safe opportunity for social interaction (Faulkner & Sparkes, 1999). Rose Hill in Holly, Michigan, for example, offers residents a wide range of options for physical activity, including yoga, aerobics, volleyball, and basketball (http://www.gouldfarm.org/). The location of Spring Lake Ranch Therapeutic Community in Cuttingsville, Vermont, provides residents with opportunities for canoeing, swimming, snowshoeing, cross-country and downhill skiing, and camping. Residents can also participate in yoga, table tennis, or billiards (http://www.springlakeranch.org/ treatment-programs/ranch-program).

Spiritual and Religious Activities

Attention to individuals' spiritual and religious beliefs and practices is important for a number of reasons. First, individuals may model their behavior based on their understandings of the precepts that underlie their religious affiliation or spiritual beliefs (Lindgren & Coursey, 1995; Mitchell & Romans, 2002; Sullivan, 1998). Spiritual beliefs have been said to constitute

> the central philosophy of life which guides people's conduct and is the core of individual existence that integrates and transcends the physical, emotional, intellectual, more-ethical, volitional, and social dimensions. (Kaye & Raghavan, 2002, p. 231)

Second, existing research findings indicate that religious/spiritual beliefs and practices often provide individuals in recovery from mental illness with a sense of meaning and control and may also serve as a source of hope, comfort, support, and acceptance (Fallot, 1997, 1998; Lindgren & Coursey, 1995; Phillips, Lakin, & Pargament, 2002; Sullivan, 1998; Tepper, Rogers, Coleman, & Malony, 2001). Third, an understanding of client beliefs and practices, and the nonjudgmental provision of opportunities for discussion, may be important to intervene in situations in which client adherence to a recommended medication and/or behavioral regimen is threatened or compromised as the result of differing or contradictory counsel that they have received from their religious authorities (Mitchell & Romans, 2002). Finally, it is important to understand a client's religious and spiritual beliefs and practices in order to distinguish beliefs that are religious in nature and those that reflect underlying pathology, such as auditory and/or visual hallucinations that are religious in nature (Loue & Sajatovic, 2008).

Individuals may experience both religion and spirituality in the context of a specific activity. However, individuals who are not engaged with a faith community may find it difficult to connect to a sense of spirituality in the context of religious practice. It may be important to help residents explore the difference(s) between religion and spirituality so that they can identify what works best for them. Religion is often seen as comprising the beliefs, ethical codes, and worship practices that unite an individual with a moral community (Joseph, 1998); in other words, religion is communal. In contrast, spirituality is individual and personal (Titone, 1991).

Some therapeutic farms arrange for clients to attend worship services of their choosing at a nearby place of worship. They may also provide opportunities for residents to participate in diverse activities that focus on spirituality. These may include yoga sessions, meditation practice, nature walks, and/or discussions relating to values.

It appears that a few therapeutic farms provide access to yoga, which may offer residents a variety of health-related benefits. Although there are various forms of yoga, standard components across these various systems include specific postures (*asanas*), breathing exercises (*pranayamas*), and meditation (*dhyana*) (da Silva, Ravindran, & Ravindran, 2009; Kirkwood, Rampes, Tuffrey, Richardson, & Pilkington, 2005; Pilkington, Kirkwood, Rampes, & Richardson, 2005). Yoga practice modulates autonomic nervous tone, reduces sympathetic tone, and activates antagonistic neuromuscular systems, leading to an increase in the relaxation response and stimulation of the limbic system (Riley, 2004). Depending upon the extent to which each of yoga's elements is emphasized, yoga may provide an opportunity for residents to engage in exercise through the stretching and the poses and to engage in spiritual reflection through the meditation. Reviews of the scientific literature relating to the use of yoga in the treatment of mental illness suggest that yoga is beneficial as an augmentation to medication in the treatment of depressive disorders (Pilkington et al., 2005), including mild to moderate major depression and dysthymia (da Silva et al., 2009), and obsessive compulsive disorder (Kirkwood et al., 2005). Yoga has also been found to provide some benefit to individuals diagnosed with schizophrenia (Duraiswamy, Thirthalli, Nagendra, & Gangadhar, 2007; Visceglia, 2007).

References

Allgower, A., Wardle, J., & Steptoe, A. (2001). Depressive symptoms, social support, and personal health behaviors in young men and women. *Health Psychology, 20*, 223–227.

American Art Therapy Association. (2013). What is art therapy? http://www.arttherapy.org/upload/whatisarttherapy.pdf. Accessed 13 Mar 2015.

Amman, R. (1991). *Healing and transformation in sandplay: Creative processes become visible.* Peru, IL: Open Court Publishing.

Amman, R. (1994). The sandtray as a garden of the soul. *Journal of Sandplay Therapy, 4*(1), 46–65.

Animal Assisted Intervention International. (2013). Glossary of terms: Animal assisted activities (AAA). http://www.animalassistedintervention.org/AnimalAssistedIntervention/Glossary/A/AnimalAssistedActivities.aspx. Accessed 18 Feb 2015.

Barker, S., & Dawson, K. (1998). The effects of animal-assisted therapy on anxiety ratings of hospitalized psychiatric patients. *Psychiatric Services, 49*, 797–801.

Barrowclough, C., Haddock, G., Tarrier, N., Lewis, S. W., Moring, J., O'Brien, R., et al. (2001). Randomized controlled trial of motivational interviewing, cognitive behavior therapy, and family interventions for patients with comorbid schizophrenia and substance use disorders. *American Journal of Psychiatry, 158*(10), 1706–1713.

Bartsch, D. A., Shern, D. L., Feinberg, L. E., Fuller, B. B., & Willett, A. B. (1990). Screening CMHC outpatients for physical illness. *Hospital and Community Psychiatry, 41*, 786–790.

Baun, M. M., Bergstrom, N., Langston, N. F., & Thomas, L. (1983). Physiological effects of human/companion animal bonding. *Nursing Research, 33*(3), 126–130.

Beck, A. M. (2005). Review of pets and our mental health: The why, the what, and the how. *Anthrozoos, 18*(4), 441–443.

Berget, B., Ekeberg, O., & Brastad, B. (2008). Animal-assisted therapy with farm animals for persons with psychiatric disorders: Effects on self-efficacy, coping ability, and quality of life: A randomized controlled trial. *Clinical Practice and Epidemiology in Mental Health, BioMed Central, 4*(9). http://www.cpementalhealth.com/content/4/1. Accessed 15 Feb 2015.

Berlyne, D. E., & Borsa, D. M. (1968). Uncertainty and the orientation reaction. *Perception and Psychophysiology, 3*, 777–779.

Berren, M. R., Hill, K. R., Merikle, E., Gonzales, N., & Santiago, J. (1994). Serious mental illness and mortality rates. *Hospital and Community Psychiatry, 45*, 604–605.

Bhui, K., & Fletcher, A. (2000). Common mood and anxiety states: Gender differences in the protective effect of physical activity. *Social Psychiatry & Psychiatric Epidemiology, 35*(1), 28–35.

Bizub, A. L., Joy, A., & Davidson, L. (2003). "It's like being in another world": Demonstrating the benefits of therapeutic horseback riding for individuals with psychiatric disability. *Psychiatric Rehabilitation Journal, 26*, 377–384.

Boik, B. L., & Goodwin, E. A. (2000). *Sandplay therapy: A step-by-step manual for psychotherapists of diverse orientations.* New York: W.W. Norton & Company.

Bruscia, K. S. (1998). *Defining music therapy* (2nd ed.). Gilsum, NH: Barcelona Publishers.

Bunt, L. (1988). Music therapy: An introduction. *Psychology of Music, 16*, 3–9.

Camacho, T. C., Roberts, R. E., Lazarus, N. B., Kaplan, G. A., & Cohen, R. D. (1991). Physical activity and depression: Evidence from the Alameda County study. *American Journal of Epidemiology, 134*, 220–231.

Colombo, G., Buono, M. O., Smania, K., Raviola, R., & de Leo, D. (2006). Pet therapy and institutionalized elderly: A study on 144 cognitively unimpaired subjects. *Archives of Geontology and Geriatrics, 42*(2), 207–216.

da Silva, T. L., Ravindran, L. N., & Ravindran, A. V. (2009). Yoga in the treatment of mood and anxiety disorders: A review. *Asian Journal of Psychiatry, 2*, 6–16.

Daykin, N., Byrne, E., Soteriou, T., & O'Connor, S. (2008). The impact of art, design and environment in mental healthcare: A systematic review of the literature. *Journal of the Royal Society for the Promotion of Health, 128*(2), 85–94.

De Domenico, G. S. (1999). The legacy of Margaret Lowenfeld: The Lowenfeld World Technique and Lowenfeld sandplay. *Sandtray Network Journal.* http://www.gapt.org/pdf_files/ARTICLES%20VOL%201-9/05-01%20THE%20LEGACY%20OF%20MARGARET%20LOWENFELD-DEDOMENICO.pdf. Accessed 13 Sept 2013.

DeRubeis, R. J., Hollon, S. D., Amsterdam, J. D., Shelton, R. C., Young, P. R., Salomon, R. M., et al. (2005). Cognitive therapy vs medications in the treatment of moderate to severe depression. *Archives of General Psychiatry, 62*, 409–416.

Dimeo, F., Bauer, M., Varahram, I., Proest, G., & Halter, U. (2001). Benefits from aerobic exercise in patients with major depression: A pilot study. *British Journal of Sports Medicine, 35*, 114–117.

Dunn, A. L., Trivedi, M. H., Kampert, J. B., Clark, C. G., & Chambliss, H. O. (2005). Exercise treatment for depression: Efficacy and dose response. *American Journal of Preventive Medicine, 28*(1), 1–8.

Dunn, A. L., Trivedi, M. H., & O'Neal, H. A. (2001). Physical activity dose-response effects on outcomes of depression and anxiety. *Medicine & Science in Sports & Exercise, 33,* S587–S597.

Duraiswamy, G., Thirthalli, J., Nagendra, H. R., & Gangadhar, B. N. (2007). Yoga therapy as an add-on treatment in the management of patients with schizophrenia: A randomized controlled trial. *ACRA Psychiatrica Scandinavica, 116,* 226–232.

Edwards, M. (1978). Art therapy in Great Britain. In D. Elliott (Ed.), *The inner eye.* Oxford, UK: Museum of Modern Art.

Elings, M. (2006). People-plant interaction. In J. Hassink & M. van Dijk (Eds.), *Farming for health: Green-care farming across Europe and the United States of America* (pp. 43–55). Dordrecht, Netherlands: Springer.

Ellingson, M. (1991). A philosophy for clinical art therapy. In H. B. Landgarten & D. Lubbers (Eds.), *Adult art psychotherapy: Issues and applications* (pp. 3–20). New York: Brunner/Mazel.

Fallot, R. D. (1997). Spirituality in trauma recovery. In M. Harris (Ed.), *Sexual abuse in the lives of women diagnosed with serious mental illness* (pp. 337–355). Amsterdam: Harwood Academic Publishers.

Fallot, R. D. (1998). Spiritual and religious dimensions of mental illness recovery narratives. In R. D. Fallot (Ed.), *Spirituality and religion in recovery from mental illness* (pp. 35–44). San Francisco, CA: Jossey-Bass Publishers.

Faulkner, G., & Biddle, S. (1999). Exercise as an adjunct treatment for schizophrenia: A review of the literature. *Journal of Mental Health, 8,* 441–457.

Faulkner, G., & Sparkes, A. (1999). Exercise as therapy for schizophrenia: An ethnographic study. *Journal of Sport and Exercise Psychology, 21,* 52–69.

Felthous, A. R., & Kellert, S. R. (1987). Childhood cruelty to animals and later aggression against people: A review. *American Journal of Psychiatry, 144,* 710–717.

Filan, S., & Llewelltn-Jones, R. (2006). Animal assisted therapy for dementia: A review of the literature. *International Psychogeriatrics, 18*(4), 597–611.

Fjeld, T., Veiersted, B., & Sandvik, L. (2002). The effect of indoor foliage plants on health and discomfort symptoms among office workers. *Indoor & Built Environment, 7,* 204–206.

Framer, M. E., Locke, B. Z., Moscicki, E. K., Dannenberg, A. L., Larson, D. B., & Radloff, L. S. (1988). Physical activity and depressive symptoms: The NHANES I epidemiology follow-up study. *American Journal of Epidemiology, 128,* 1340–1351.

Friedmann, E., Katcher, A. H., Lynch, J. J., & Thomas, S. A. (1980). Animal companions and one-year survival of patients after discharge from a coronary care unit. *Public Health Reports, 95*(4), 307–312.

Galper, D. I., Trivedi, M., Barlow, C. E., Dunn, A. L., & Kampert, J. B. (2006). Inverse association between physical inactivity and mental health in men and women. *Medicine & Science in Sports & Exercise, 38*(1), 173–178.

Gold, C., Heldal, T. O., Dahle, T., & Wigram, T. (2005). Music therapy for schizophrenia or schizophrenia-like illnesses. *Cochrane Database Systems Review, 2,* CD004025.

Gonzalez, M. T., Hartig, T., Patil, G. G., & Martinsen, E. W. (2011). A prospective study of existential issues in therapeutic horticulture for clinical depression. *Issues in Mental Health Nursing, 32,* 73–81.

Goodwin, R. D. (2003). Association between physical activity and mental disorders among adults in the United States. *Preventive Medicine, 36,* 698–703.

GrowthPoint. (1999). Your future starts here: Practitioners determine the way ahead. *Growth Point, 79,* 4–5.

Hacking, S., Secker, J., Spandler, H., Kent, L., & Shenton, J. (2008). Evaluating the impact of participatory art projects for people with mental health needs. *Human and Social Care in the Community, 16*(6), 638–648.

Hafen, M., Rush, B., Reisbig, A., McDaniel, K., & White, M. (2007). The role of family therapists in veterinary medicine: Opportunities for clinical services, education, and research. *Journal of Marital and Family Therapy, 33*(2), 165–176.

Hamer, M., Stamatakis, E., & Steptoe, A. (2009). Dose-response relationship between physical activity and mental health: The Scottish Health Survey. *British Journal of Sports Medicine, 43,* 1111–1114.

Harper, F. D., & Bruce-Sanford, G. C. (1989). *Counseling techniques: An outline and overview.* Alexandria, VA: Douglass Publishers.

Hassink, J. (2002). Combining agricultural production and care for persons with disabilities: A new role of agriculture and farm animals. In A. Cirstovao & L. O. Zorini (Eds.), *Farming and rural systems research and extension. Local identities and globalization* (pp. 332–341). Fifth IFSA European symposium, Florence, 8 Apr 2002. Florence: Agenzia Regionale per lo Sviluppo e l'Innovazione nel Settore Agro-forestale della Regione Toscane (ARSIA). http:// ifsa.boku.ac.at/cms/fileadmin/Proceeding2002/2002_WS04_02_Hassink.pdf. Accessed 18 Feb 2015.

Hayashi, N., Tanabe, Y., Nakagawa, S., Noguchi, M., Iwata, C., Koubuchi, Y., et al. (2002). Effects of group musical therapy on inpatients with chronic psychoses: A controlled study. *Psychiatry and Clinical Neurosciences, 56,* 187–193.

Hernandez-Peon, R. (1961). The efferent control of afferent signals entering the central nervous system. *Annals of the New York Academy of Science, 89,* 866–882.

Hertfordshire Partnership, NHS Foundation Trust. (n.d.). *Cognitive behavioural therapy skills training workbook.* http://www.hpft.nhs.uk/_uploads/documents/help-for-adults/cbt-workshop-booklet_web.pdf. Accessed 18 Mar 2015.

Hutton, D. (2004). Margaret Lowenfeld's "World Technique". *Clinical Child Psychology and Psychiatry, 9,* 605–612.

Jessen, J., Cardiello, F., & Baun, M. (1996). Avian companionship in alleviation of depression, loneliness, and low morale of older adults in skilled rehabilitation units. *Psychological Reports, 78*(2), 339–348.

Jorgenson, J. (1997). Therapeutic use of companion animals in health care. *Journal of Nursing Scholarship, 29*(3), 249–254.

Joseph, M. V. (1998). Religion and social work practice. *Social Casework: Journal of Contemporary Social Work, 69,* 443–452.

Jung, C. G. (1928). *Self and unconsciousness.* (trans. R.F.C. Hull). Bollingen Series XX. Princeton, NJ: Princeton University Press.

Kam, M. C. Y., & Siu, A. M. H. (2010). Evaluation of a horticultural activity programme for persons with psychiatric illness. *Hong Kong Journal of Occupational Therapy, 20*(2), 80–86.

Kaplan, S. (1992). The restorative environment: Nature and human experience. In D. Relf (Ed.), *The role of horticulture in human well-being and social development: A national symposium 19–21 Apr 1990, Arlington, Virginia* (pp. 134–142). Portland, OR: Timber Press.

Kaye, J., & Raghavan, S. K. (2002). Spirituality in disability and illness. *Journal of Religion and Health, 41*(3), 231–242.

Kirkwood, G., Rampes, H., Tuffrey, V., Richardson, J., & Pilkington, K. (2005). Yoga for anxiety: A systematic review of the research evidence. *British Journal of Sports Medicine, 39,* 884–891.

Koran, L. M., Sox, H. C., Morton, K. L., Moltzen, S., Sox, C. H., Kraemer, H. C., et al. (1989). Medical evaluation of psychiatric patients. *Archives of General Psychiatry, 36,* 414–447.

Kramer, E. (2000). In L. A. Gerity (Ed.), *Art as therapy: Collected papers.* London: Jessica Kingsley Publishers.

Kruger, K. A., & Serpell, J. A. (2006). Animal-assisted interventions in mental health: Definitions and theoretical foundations. In A. H. Fine (Ed.), *Handbook on animal-assisted therapy: Theoretical foundations and guidelines for practice* (2nd ed., pp. 21–38). San Diego, CA: Academic Press.

Kruger, K. A., Trachtenberg, S. W., & Serpell, J. A. (2004, July). *Can animals help humans heal?: Animal-assisted interventions in adolescent mental health.* Philadelphia: Center for the Interaction of Animals and Society, University of Pennsylvania School of Veterinary Medicine. http://research.vet.upenn.edu/Portals/36/media/CIAS_AAI_white_paper.pdf. Accessed 18 Feb 2015.

Lawrence, D. M., Holman, C. D., Jablensky, A. V., & Hobbs, M. S. (2003). Death rate from ischaemic heart disease in Western Australian psychiatric patients 1980–1998. *British Journal of Psychiatry, 182,* 31–36.

Lean, M. E., & Pajonk, F. G. (2003). Patients on atypical antipsychotic drugs: Another high-risk group for type 2 diabetes. *Diabetes Care, 26,* 1597–1605.

Levinson, B. (1970). Pets, child development, and mental illness. *Journal of the American Veterinary Medicine Association, 157*(11), 1759–1766.

Levinson, B. (1982). The future of research into the relationship between people and their companion animals. *International Journal for the Study of Animal Problems, 9,* 283–294.

Lindgren, K. N., & Coursey, R. D. (1995). Spirituality and mental illness: A two-part study. *Psychosocial Rehabilitation Journal, 18*(3), 93–111.

Lloyd, C., Wong, S. R., & Petchkovsky, L. (2007). Art and recovery in mental health: A qualitative investigation. *British Journal of Occupational Therapy, 70*(5), 207–214.

Loue, S., Mendez, N., & Sajatovic, M. (2008). Preliminary evidence for the integration of music into HIV prevention for severely mentally ill Latinas. *Journal of Immigrant and Minority Health, 10*(6), 489–495.

Loue, S., & Sajatovic, M. (2008). Auditory and visual hallucinations in a sample of severely mentally ill Puerto Rican women: An examination of the cultural context. *Mental Health, Culture, and Religion, 11*(6), 597–608.

Lyddiatt, E. M. (1972). *Spontaneous painting and modelling.* New York: St. Martin's Press.

Lynch, T. R., Morse, J. Q., Mendelson, T., & Robins, C. J. (2003). Dialectical behavior therapy for depressed older adults: A randomized pilot study. *American Journal of Geriatric Psychiatry, 11,* 33–45.

Mallon, G. P. (1994). Cow as co-therapist: Utilization of farm animals as therapeutic aides with children in residential treatment. *Child and Adolescent Social Work Journal, 11*(6), 455–474.

Marteniuk, R. G. (1976). *Information processing in motor skills.* New York: Holt, Rinehart, & Winston.

Martinsen, E. W. (1994). Physical activity and depression: Clinical experience. *Acta Psychiatrica Scandinavica, 89*(Suppl. S377), 23–27.

Martinsen, E. W., Hofart, A., & Solberg, O. (1989). Aerobic and non-aerobic forms of exercise in the treatment of anxiety disorders. *Stress Medicine, 5*(2), 115–120.

Miller, A. (1998). The role of community arts in psychosocial rehabilitation. In *VICSERV conference papers* (pp. 167–170). Melbourne: Psychiatric Disability Services of Victoria.

Mitchell, L., & Romans, S. (2002). Spiritual beliefs in bipolar affective disorder: Their relevance for illness management. *Journal of Affective Disorders, 75,* 247–257.

Molloy, T. (1997). Art psychotherapy and psychiatric rehabilitation. In K. Killick & J. Schaverien (Eds.), *Art, psychotherapy and psychosis* (pp. 237–259). London: Routledge.

Morter, S. (1997). Where words fail: A meeting place. In K. Killick & J. Schaverien (Eds.), *Art, psychotherapy and psychosis* (pp. 219–236). London: Routledge.

National Association of Cognitive-Behavioral Therapists. (2014). Cognitive-behavioral therapy. http://www.nacbt.org/whatiscbt.htm. Accessed 18 Mar 2015.

Phillips, R. E., III, Lakin, R., & Pargament, K. I. (2002). Development and implementation of a spiritual issues psychoeducational group for those with serious mental illness. *Community Mental Health Journal, 8*(6), 487–495.

Piette, J., Richardson, C., & Valenstein, M. (2004). Addressing the needs of patients with multiple chronic illnesses: The case of diabetes and depression. *American Journal of Managed Care, 10,* 41–51.

Pilkington, K., Kirkwood, G., Rampes, H., & Richardson, J. (2005). Yoga for depression: The research evidence. *Journal of Affective Disorders, 89*, 13–24.

Professional Association of Therapeutic Horsemanship International. (2013). EAAT definitions. http://www.pathintl.org/component/content/article/27-resources/general/193-eaat-definitions. Accessed 18 Feb 2015.

Professional Association of Therapeutic Horsemanship International. (2015). Learning about therapeutic riding. http://www.pathintl.org/component/content/article/27-resources/general/198-learn-about-therapeutic-riding. Accessed 18 Feb 2015.

Quindlen, A. (2007). *Good dog, stay!* New York: Random House.

Rajski, P. (2003). Finding God in the silence: Contemplative prayer and therapy. *Journal of Religion and Health, 42*(3), 181–190.

Rector, N. A., & Beck, A. T. (2001). Cognitive behavioral therapy for schizophrenia: A review. *Journal of Nervous and Mental Disease, 189*(5), 278–287.

Rehor, P. R., Dunnagan, T., Stewart, C., & Cooley, D. (2001). Alteration of mood state after a single bout of noncompetitive and competitive exercise programs. *Perceptual and Motor Skills, 93*, 249–256.

Rigdon, J. D., & Tapia, F. (1977). Children who are cruel to animals—A follow-up study. *Journal of Operational Psychiatry, 8*(1), 27–36.

Riley, D. (2004). Hatha yoga and the treatment of illness. *Alternative Therapies in Health & Medicine, 10*(2), 20–21.

Rynearson, E. K. (1978). Humans and pet attachment. *British Journal of Psychiatry, 133*, 550–555.

Schmitz, N., Kruse, J., & Kugler, J. (2004). The association between physical exercises and health-related quality of life in subjects with mental disorders: Results from a cross-sectional study. *Preventive Medicine, 39*, 1200–1207.

Scott, J., Paykel, E., Morriss, R., Bentall, R., Kinderman, P., Johnson, T., et al. (2006). Cognitive-behavioural therapy for severe and recurrent bipolar disorders. *British Journal of Psychiatry, 188*, 313–328.

Sempik, J., & Aldridge, J. (2006). Care farms and care gardens: Horticulture as therapy in the UK. In J. Hassink & M. van Dijk (Eds.), *Farming for health: Green-care farming across Europe and the United States of America* (pp. 147–161). Dordrecht, Netherlands: Springer.

Sernyak, M. J., Leslie, D. L., Alarcon, R. D., Losonczy, M. F., & Rosenheck, E. (2002). Association of diabetes mellitus with use of atypical neuroleptics in the treatment of schizophrenia. *American Journal of Psychiatry, 159*, 561–566.

Serpell, J. A. (1990, April). Evidence for long term effects of pet ownership on human health. *Waltham symposium 20: Pets, benefits & practice. First European congress of the British Small Animal Veterinary Association*. Cheltenham, England: BVA Publications.

Siegel, J. M., Angulo, F. J., Detels, R., Wech, J., & Mullen, A. (1999). AIDS diagnosis and depression in the Multicenter AIDS Cohort Study: The ameliorating impact of pet ownership. *AIDS Care, 11*(2), 157–170.

Skailes, C. (1997). The forgotten people. In K. Killick & J. Schaverien (Eds.), *Art, psychotherapy and psychosis* (pp. 198–218). London: Routledge.

Stephens, T. (1988). Physical activity and mental health in the United States and Canada: Evidence from four population surveys. *Preventive Medicine, 17*, 35–47.

Strawbridge, W. J., Deleger, S., Roberts, R. E., & Kaplan, G. A. (2002). Physical activity reduces the risk of subsequent depression in older adults. *American Journal of Epidemiology, 156*, 328–334.

Sullivan, W. P. (1998). Recoiling, regrouping, and recovering: First-person accounts of the role of spirituality in the course of serious mental illness. *New Directions for Mental Health Services, 80*, 25–33.

Talwar, N., Crawford, M. J., Maratos, A., Nur, U., McDermott, O., & Procter, S. (2006). Music therapy for in-patients with schizophrenia: Exploratory randomized controlled trial. *British Journal of Psychiatry, 189*, 405–409.

Tang, W., Yao, X., & Zheng, Z. (1994). Rehabilitative effect of music therapy for residual schizo-phrenia: A one-month randomized controlled trial in Shanghai. *British Journal of Psychiatry*, (Suppl. 24), 38–44.

Tapia, F. (1971). Children who are cruel to animals. *Child Psychiatry and Human Development*, 2(2), 70–77.

Tepper, L., Rogers, S. A., Coleman, E. M., & Malony, H. N. (2001). The prevalence of religious coping among persons with persistent mental illness. *Psychiatric Services, 52*(5), 660–665.

Thaut, M. H. (1990). Neuropsychological processes in music perception and their relevance in music therapy. In R. F. Unkefer (Ed.), *Music therapy in the treatment of adults with mental disorders* (pp. 3–32). New York: Schirmer Books.

Thomson, C. (1981). Variations on a theme by Lowenfeld: Sandplay in focus. In C.G. Jung Institute of San Francisco (Ed.), *Sandplay studies: Origins, theory, and practice* (pp. 5–20). Boston: Sigo Press.

Titone, A. M. (1991). Spirituality and psychotherapy in social work practice. *Spirituality & Social Work Communicator, 2*, 7–9.

Ulman, E. (1975). Art therapy: Problems of definition. In E. Ulman & P. Dachinger (Eds.), *Art therapy in theory and practice* (pp. 3–13). New York: Schocken Books.

Ulrich, R. S., & Parsons, R. (1992). Influences of passive experiences with plants on individual well-being and health. In D. Relf (Ed.), *The role of horticulture in human well-being and social development: A national symposium, 19–21 Apr 1990, Arlington, Virginia* (pp. 93–105). Portland, OR: Timber Press.

Unruh, A. M. (2004). The meaning of gardens and gardening in daily life: A comparison between gardeners with serious health problems and healthy participants. *Acta Horticulturae, 639*, 67–73.

Van Gool, C. H., Kempen, G. I., Penninx, B. W., Deeg, D. J., Beekman, A. T., & Van Euk, J. T. (2003). Relationship between changes in depressive symptoms and unhealthy lifestyles in late middle aged and older persons: Results from the Longitudinal Aging Study Amsterdam. *Age & Aging, 32*, 81–87.

Visceglia, E. (2007). Healing mind and body: Using therapeutic yoga in the treatment of schizo-phrenia. *International Journal of Yoga Therapy, 17*, 95–103.

Walsh, F. (2009). Human-animal bonds I: The relational significance of companion animals. *Family Process, 48*(4), 462–480.

Wassertheil-Smoller, S., Shumaker, S., Ockene, J., Talavera, G. A., Greenland, P., Cochrane, B., et al. (2004). Depression and cardiovascular sequelae in postmenopausal women: The Women's Health Initiative. *Archives of Internal Medicine, 164*, 289–298.

Weinrib, E. L. (1983). *Images of the self*. Boston, MA: Sigo Press.

Wiesinger, G., Neuhauser, F., & Putz, M. (2006). Farming for health in Austria: Farms, horticul-tural therapy, animal-assisted therapy. In J. Hassink & M. van Dijk (Eds.), *Farming for health: Green-care farming across Europe and the United States of America* (pp. 233–248). Dordrecht, Netherlands: Springer.

Wigram, T. (2000). Theory of music therapy: Lectures to undergraduate and post-graduate stu-dents, University of Aalborg (unpublished communication). Cited in Wigram, T., Pedersen, I.M., & Bonde, L.O. (Eds.). (2002). *A comprehensive guide to music therapy: Theory, clinical practice, research and training*. London: Jessica kingsley Publishers.

Wigram, T., Pedersen, I. N., & Bonde, L. O. (2002). *A comprehensive guide to music therapy: Theory, clinical practice, research and training*. London: Jessica Kingsley Publishers.

Wood, C. (1997). The history of art therapy and psychosis (1938–95). In K. Killick & J. Schaverien (Eds.), *Art, psychotherapy and psychosis* (pp. 145–175). London: Routledge.

Wyshak, G. (2001). Women's college physical activity and self-reports of physician-diagnosed depression and of current symptoms of psychological distress. *Journal of Women's Health & Gender Based Medicine, 10*, 363–370.

Zasloff, R. L., & Kidd, A. H. (1994). Loneliness and pet ownership among single women. *Psychological Reports, 75*(2), 747–752.

Chapter 3
Approaches to Organizational Structure and Financing

This chapter provides an overview of the diverse approaches utilized on organizing and sustaining care farms in Europe and therapeutic farms in the United States. The legal structure of a care or therapeutic farm and its approach to financing often depend upon the options that are available within the legal system that governs its existence and the extent to which government funding is or is not available for mental health services generally and for care on such farms specifically.

Organizing and Financing Care Farms in Europe

The Impetus for the Development of Care Farms

The impetus for the establishment of care farms arises from two concurrent but independent developments: the diversification of farms to include non-farming activities and the movement of many European countries from the provision of care to mentally ill individuals in residential institutions to a more community-based model.

European agriculture has undergone extensive change since the 1960s in an effort to address the consequences of and challenges resulting from increased specialization and capital-intensive production systems: climbing costs, declining profits, and a reduction in farmers' social integration (Hine, Peacock, & Pretty, 2008). Efforts to engage in multifunctional agricultural activities—non-farming activities conducted on a farm—have variously led to the establishment of agro-tourism undertakings, landscape management services, and the care and/or employment of vulnerable individuals (Dessein, 2008, pp. 13–15).

The European health-care and social service sectors have also undergone an evolution. Earlier approaches to the provision of services to individuals with mental illness were often premised on and rigidly adhered to a biomedical model of mental

© Springer International Publishing Switzerland 2016
S. Loue, *Therapeutic Farms*, SpringerBriefs in Social Work,
DOI 10.1007/978-3-319-13539-7_3

illness. In contrast, increasing emphasis is now placed on the integration and empowerment of consumers of mental health-related services (Barnes, 2008, p. 30; Dessein, 2008, p. 15; McGloin, 2009, p. 106).

The concurrent need to diversify agriculture and both to empower and to reintegrate those with mental illness into their communities led to the conceptualization of "farming for health." The Community of Practice Farming for Health has defined farming for health as being based on a combination of agriculture and care. The focus is both on the farming system (which includes components as the farm enterprise itself, operational management, the farmer and farmer's social environment) and the care sector (including, e.g., the help-seeker, the institution, and the care professional). The result is a very diverse picture of care-seekers involved in on-farm activities (Dessein, 2008, pp. 15–16).

Social farming, which represents one type of farming for health, consists of farming practices aimed at promoting disadvantaged people's rehabilitation, education, and care and/or toward the integration of people with "low contractual capacity" (i.e., intellectual and physical disabilities, convicts, those with drug addiction, minors, migrants) but also practices that support services in rural areas for specific target groups such as children and the elderly (Di Iacovo & O'Connor, 2009, p. 11).

Because the impetus for the establishment of a care farm may derive from either a perceived need to diversify agricultural activities or a desire to modify the delivery of mental health services, either of two different pathways may lead to the establishment of a care farm. One pathway results from the decision of a farmer on an existing farm to diversify the farm-based activity and sources of revenue (Hassink, Hulsink, & Grin, 2012, p. 25). In the second scenario, a desire to engage in nonagricultural health-related activities prompts interest in the purchase of or collaboration with an existing farm. These differing pathways have implications for the approach of care farms toward individuals in need of services, the farms' legal organization, and the financing mechanisms.

Conceptual Frameworks

As noted in Chap. 1, Dessein, Bock, and de Krom (2013) have suggested that care farms operate from a perspective that emphasizes public health, social inclusion, or multifunctional agriculture. To review, the public health frame emphasizes the potential health benefits that can be derived from the provision of physical and spiritual experiences in a natural setting that encompasses seasonal cycles (De Bruin et al., 2010; De Vires, 2006). The social inclusion framework, which recognizes that persons with mental illness have been excluded from the larger society, focuses on (1) the reintegration of excluded persons into society through activities formulated to increase their knowledge and skills, (2) the reestablishment of their ability to engage in work, and (3) the development of their self-esteem (Dessein et al., 2013, p. 53). Participation in agricultural labor is seen as a mechanism for the restoration of a structured routine and the promotion of interactive activities. The public health frame appears to predominate in Germany, Austria, and the United Kingdom, the social inclusion framework in Ireland and Italy, and the multifunctional agriculture approach in Belgium, the Netherlands, Norway, and Slovenia (cf. Dessein et al., 2013, pp. 53–55).

Friedel, Mathijs, and Van Molle (2010, p. 34) have suggested that the operational frames of care farms can be delineated more finely into five, rather than three, categories. According to this typology, the agricultural frame is premised on the need of the farmer to diversify his or her income in order to survive. The welfare frame, reflecting the "Dutch model," focuses on social farming as a mechanism by which to increase the efficiency with which care is provided and simultaneously reduce the costs associated with that care. The social economy frame, which Friedel and colleagues refer to as the "Italian model," views social farming as a means by which to provide members of vulnerable groups with employment opportunities. The regional frame views social farming as an opportunity to develop community-based initiatives and to forge local partnerships. Lastly, the quality of care frame emphasizes the delivery of better quality care to service users through social farming.

Organizational Models

Care farms in the Netherlands, and perhaps other Western European countries as well, may reflect any one of ten organizational models as a function of three key dimensions: organizational structure; strategy, including the motivation that prompted the establishment of the farm; and environment (Hassink et al., 2012, p. 9). These include the following:

1. *Helping hand alliances*, which focus on agricultural production and maintain collaborative relationship with an institution. Care farms utilizing this model are initiated jointly by a care institution and a farmer. These farms are generally small-scale commercial enterprises. Employees of the care institution coach the farmer on the care services (Hassink et al., 2012, p. 16).
2. *Independent low-care farms*, which focus on agricultural production, are relatively independent from any care institution and are often family owned. Care services are provided, on average, to fewer than six clients per day, and the revenue from care services generally constitutes less than 25 % of the farm's revenue. This model accounts for 15 % of the care farms in the Netherlands (Hassink et al., 2012, p. 16).
3. *Independent integrated care farms* account for approximately 20 % of the Netherlands' care farms. They are relatively independent from formal care institutions and maintain a focus on both agricultural production and the provision of care services. Clients often number between 7 and 15 daily and the income derived from the provision of these care services is approximately 25 % of the farm's revenue (Hassink et al., 2012, pp. 16–17).

4. *Independent care focus farms* provide care services, but maintain independence from formal care institutions. They are similar to independent integrated farms except that agricultural activity is more limited and care services more predominant. More than three-quarters of the farm income is attributable to revenue received from the provision of the care services (Hassink et al., 2012, p. 17).

5. *Care focus alliances* emphasize the provision of care services while maintaining a close relationship with a formal care institution. The care farm is often established on an existing farm; the collaboration with a care institution reduces the potential financial risks and burdens associated with the provision of the care services. Income derived from the provision of care services typically accounts for more than three-quarters of the farm's income. Fewer than 5 % of the Netherlands' care farms are of this nature (Hassink et al., 2012, p. 18).

6. *Independent added care farms* have been established by former employees of the care sector who establish collaborations with the owner of an existing farm. In this model, the former care employee is responsible for providing the care services, while the farmer benefits from the clients' labor on the farm. Services are generally provided for no more than seven clients. Approximately 5 % of the Netherlands' care farms fall into this category (Hassink et al., 2012, pp. 18–19).

7. *Independent care focus farms (former employee)* are similar to the independent care focus farms in that their focus is on the provision of care services, agricultural production is limited, and care services account for more than 75 % of the farms' revenue. They are distinguished from independent care focus farms in that they are established by former employees of the care sector. This model accounts for about 10 % of the Netherlands' care farms (Hassink et al., 2012, p. 19).

8. *Care focus farm alliances* emphasize the provision of care services and maintain a collaborative relationship with an institution. These farms are initiated in collaboration with an existing care institution, rather than through a collaboration with an existing farm or by purchasing a farm. Some such care farms may have received their initial support from governmental care innovation funds and through partnerships with existing care institutions. In comparison with the previously noted models of care farms, these farms provide services to a greater number of clients, averaging 12–15 clients per day. Less than 1 % of the care farms in the Netherlands can be characterized as care focus farm alliances (Hassink et al., 2012, pp. 19–20).

9. *Independent living-working communities* were established during the 1970s and 1980s with a focus on care services while maintaining their independence from a care institution. They have been described as being "rooted in the societal changes in the 1960s and are part of a subculture that opposed materialism, authority, and exploitation of earth and mankind" (Hassink et al., 2012, p. 20). These farms are living, working communities that have been accredited to provide care to individuals with psychiatric diagnoses and intellectual disabilities. Over time, they have become increasingly professionalized with the hiring of a director and the establishment of a supervisory board. Many of these farms provide services to more than 60 clients a day. Such farms account for about 3 % of the Netherlands' care farms.

10. *Independent care focus farms (other)* are often established on an existing farm in the local area by the families of the individuals who will be the farm's clients, and accordingly, their emphasis is on the provision of care services. It is estimated that 10 % of the Netherlands' care farms fall into this category (Hassink et al., 2012, p. 21).

Approaches to Financing

Three primary approaches to the financing of care farms have been identified: support from a governmental entity and/or a mental health-care organization, revenue obtained from the production of the farm clients and others working on the farm, and direct payment from the clients and/or collaborating institutions (Dessein et al., 2013; Friedel et al., 2010; Iancu, Zweekhorts, Veltman, van Balkom, & Bunders, 2015). Care farms receiving support from a mental health-care organization may be owned by the health-care organization or may be a private care farm that is working in collaboration with a mental health-care organization (Iancu et al., 2015, p. 176). (Despite the diversity of organizational structures utilized by care farms, some scholars do not recognize farms engaged in the provision of care as care farms unless "there is a money flow from agricultural production as well as from care activities. When both money flows are present and there is a connection between both activities we speak of a care farm" (Roest, Oosting, Ferwerda-van Zonneveld, & Caron-Flinterman, 2010, p. 306; Van Schaik, 1997, p. 20)).

In the Netherlands, the collective health insurance for long-term care (the AWBZ) has been providing financial support since 1995 for clients receiving services at AWBZ-accredited institutions (Dessein et al., 2013, p. 54). In some cases, regional platforms have been established to facilitate fund flow and the placement of clients with care farms (Roest et al., 2010, p. 308). Some regional platforms have a AWBZ accreditation and receive funds from the health-care system and then pass them on to the farms for their provision of care.

Alternatively, the clients can receive funding to pay for care outside of formal health-care institutions through a personal health budget (*Persoonsgebonden Budget or PGB*) (Dessein et al., 2013, p. 55; Hassink, Zwartbol, Agricola, Elings, & Thissen, 2007, p. 22; Iancu et al., 2015, p. 176). The PGB is similar to the *direct payments* in the United Kingdom, the *prestation spécifique dépendance* in France, and the *Pflegegeld* in Germany (Iancu et al., 2015, p. 176). In either case, the client must first establish that he or she needs AWBZ-supported care. In 2004, care farming in the Netherlands was financed through direct payments from health insurance (58 %) and personal health-care budgets (34 %). Due to budget cuts, AWBZ support is now available only to individuals with serious health problems and for activities that are deemed to be medically necessary. It has been hypothesized that these budget cuts may lead to an increase in collaboration between health-care institutions and care farmers due to the lesser costs achievable through the care farms as compared to the health-care institutions (Dessein et al., 2013, p. 55). A third possibility for fund flow lies with care institutions that contract directly with the care farms to provide specified services (Roest et al., 2010, p. 308).

Care farming in Belgium provides yet another example of government support for such undertakings. Beginning in the early 2000s, the health-care sector and the Ministry of Public Health began to explore possible mechanisms through which clients could be reintegrated into society through unpaid work (de Krom & Dessein, 2013, p. 20). Farms appeared to provide one such noninstitutional environment. The Support Centre for green care was established in 2004 through the collaborative efforts of the Flemish Farmers Union, the cooperative financial group Cera, and KVLV, a rural women's group that provides care services to promote the use of the green environment together with care for vulnerable groups (de Krom & Dessein, 2013, p. 19). The following year, the Flemish government promulgated legislation that allows farmers to apply for a subsidy of € 40 per day to compensate them for the loss of agricultural production due to the provision of care to no more than three clients per day. Applicants are required to collaborate with a care facility recognized by the Ministry of Public Health or with a counseling center for high school students that falls under the Ministry of Education (de Krom & Dessein, 2013, p. 19). An official care farm contract is used to delineate specifically the responsibilities of the farmer, the care facility, and the client. The requirement of collaboration with a care facility recognized by the Ministry of Public Health was imposed in order to ensure that clients would reside at the farms for their own benefit and not as the source of cheap labor for the benefit of the farmer (de Krom & Dessein, 2013, p. 20). Researchers conducting interviews with farmers participating in the program reported that farmers valued the subsidy as a recognition of their efforts, but that it held little economic value (de Krom & Dessein, 2013, p. 21).

Challenges Associated with Care Farm Services for Persons with Mental Illness

Numerous challenges have been noted in the provision of services through care farms to persons with mental illness. Foremost among these is the lack of adequate public funding to sustain such ventures (Friedel et al., 2010). There appears to be significant disagreement regarding the desirability of having service users pay for the services received, as well as to the negotiation of contracts between government agencies and welfare organizations with individual farmers. The hesitation to pursue the former option appears to be premised on a belief that consumers with mental illness may lack the ability to spend their funds wisely. The latter possibility is disfavored due to farmers' relatively weaker bargaining position vis-à-vis governmental agencies and apprehension related to the agencies' expectations if such agreements were to be negotiated (Friedel et al., 2010, pp. 41–42).

Whether the services provided by the farmers involved in care farm enterprises should be further professionalized appears to be a highly controversial issue at some sites (Friedel et al., 2010, pp. 40–41, 49). Proponents of further professionalization suggest that such efforts would lead to a higher quality of care for the service users and increased follow-up, while others maintain that such an effort would invariably lead to increased regulation of the enterprises. Research findings suggest that the availability of mental health professionals to care farm users may, indeed, lead to an increased focus on users' interests and needs. Researchers conducting a mixed methods study of care farms in the Netherlands concluded that those care farms that had professional supervisors were more likely to focus on service users' transition to the labor market (Iancu et al., 2015, p. 183). Germany has somewhat addressed the issue of quality of care through the provision of farm-based care in collaboration with health-care institutions (Haubenhofer, Elings, Hassink, & Hine, 2010).

Additional challenges in the delivery of care through a care farm-based model may include, depending upon the country in which the farm is situated, the lack of clear legal status for the service users, who may be erroneously viewed as illegal workers; long waiting lists for placement at a suitable care farm; dumping practices, whereby individuals for whom a suitable placement cannot be found are "dumped" on care farms as a last resort; the absence of a system to monitor the quality of services provided; a lack of respect for the farmers providing such services (Friedel et al., 2010, pp. 51–52); and transportation challenges for clients participating in nonresidential care farm programs (Roest et al., 2010, p. 313).

Marketing the Care Farm Concept

Vik and Farstad (2009, p. 543) identified four constituencies of care farms: the service users, the purchasers of the services, the farmers, and the agricultural authorities. Because their research focused specifically on care farms in Norway, they identified the service purchasers as local authorities and municipalities. They noted that the sustainability of care farms depends, first, on the need for such services; second, on the willingness of an adequate number of farmers to assume this role; and, third, on the willingness of the potential buyers to pay for them. However, buyers' willingness to assume the expense of services by care farms may be reduced due to the current inability to identify which components of care farm services are more likely to yield positive outcomes for a specific class of service users, e.g., psychiatric patients (Vik & Farstad, 2009, pp. 545–546).

The Organization and Financing of US-Based Therapeutic Farms

The Impetus for the Development of Therapeutic Farms

Unlike many of Europe's care farms and green farms, US-based therapeutic farms were established specifically to provide an alternative approach to the treatment of mental illness. In some cases, acquisition of the farm was motivated by a perceived need for a farm as a venue for the provision of services. As an example, William and

Agnes Gould purchased the property on which Gould Farm rests in 1913, more than a decade after William Gould had conceived of a plan to care for persons in need of emotional rehabilitation (Gould Farm, n.d.). In other cases, a farm purchased to fulfill a nonagricultural, social service mission was revisioned to provide services for mentally ill persons. Spring Lake Ranch, for example, was originally purchased to serve as a summer camp for boys from New York City (Spring Lake Ranch Therapeutic Community, n.d.). It was later revisioned by its founders Wayne Sarcka and Elizabeth Man, to provide an alternative treatment approach for mental illness. Both of the owners had been impacted by their interactions with individuals struggling with mental illness: Sarcka had worked with shell-shocked soldiers recovering from their experiences during World War I, and Elizabeth's brother had been diagnosed with schizophrenia.

Accordingly, the therapeutic farms are firmly linked to the health-care sector rather than the agricultural sector; agricultural activity in its various forms serves as a vehicle to facilitate recovery from mental illness, rather than as an end unto itself. As a result, one does not see in the United States the national cross-sector meetings that are held in some European countries to encourage care farm business. And unlike the situation in some European countries that utilize care farms, such as Norway and the Netherlands, federal and state governmental priorities encompass neither therapeutic farms as a means to promote agricultural production nor as an alternative service modality for mental illness.

The Organization of Therapeutic Farms

The majority, if not all, of the US therapeutic farms are currently organized as charitable nonprofit organizations, a status that has both organizational and financial implications. Organizations that are classified as such under the US federal tax code are entitled to an exemption from federal income tax (Internal Revenue Code § 501(c)(3), 2014). This tax status requires that the organization be organized as a corporation, trust, or unincorporated association. Additionally, the organizing documents, such as the articles of incorporation, trust documents, or articles of association, must indicate that:

- The purpose of the organization is limited to one of the following purposes: charitable, religious, scientific, literary, fostering national or international sports competition, preventing cruelty to children or animals, and testing for public safety.
- The organization will not expressly permit activities that do not further its exempt purpose(s).
- The organization will permanently dedicate its assets to only exempt purposes (United States Department of the Treasury, 2014, pp. 3–4).

Charitable organizations are further defined; the basis of classification that is most relevant to therapeutic farms is that pertaining to the "relief of the poor, the distressed, or the underprivileged" (United States Department of the Treasury, 2014, p. 4).

Organizations that qualify for tax-exempt status under Internal Revenue Code Section 501(c)(3) are further classified as either a public charity or a private foundation. The distinction is important because donations to public charities are more likely to be tax deductible to the donor than those to private foundations; this has implications for the organization's ability to establish a broader base of financial support for its efforts. A therapeutic farm may be classifiable as a public charity if:

* It provides medical care.
* It receives a substantial part of its support in the form of contributions from publicly supported organizations, governmental units, and/or the general public.
* It normally receives not more than one-third of its support from gross investment income and more than one-third of its support from contributions, membership fees, and gross receipts related to its exempt functions (United States Department of the Treasury, 2014, pp. 5–6).

The organization of a therapeutic farm as a charitable nonprofit tax-exempt corporation under federal law does not ensure that the farm will also enjoy tax-exempt status under state law. (United States Department of the Treasury, Internal Revenue Service, 2015). The organization of the therapeutic farm as a corporation, for example, is subject to state law; state tax-exempt status is similarly dependent on state law and must be applied for independent of the application for federal tax-exempt status (Ohio Secretary of State, Business Services Division, n.d.).

Operation as a therapeutic farm requires not only a legal organization but also a license to operate from the state in which the farm is located. The nature of that license varies depending upon the services offered by a particular therapeutic farm and the categories of licenses available within that particular state. As an example, CooperRiis Healing Community, located in North Carolina, is licensed as a group home. However, because it houses more than nine residents, it must seek a waiver of the nine-person limitation each year when it applies for a renewal of its license.

Approaches to Financing

As is the case with care farms in Europe, the financing options available to therapeutic farms in the United States depend greatly on the extent to which government prioritizes and is willing to fund care for mentally ill persons in general and care furnished by therapeutic farms specifically. The government funding decisions may occur at either the state or the local level or both.

As an example, Rose Hill in Michigan derives a portion of its funding from the State of Michigan through payments made through the Medicaid insurance system. Medicaid is a joint program of the state and federal governments that assists with medical costs for persons who have limited income and resources available to them. However, the eligibility requirements for and the type and amount of benefits vary across the states (Centers for Medicare & Medicaid Services, n.d.). (See Chap. 7 for a detailed discussion of the upcoming challenges that therapeutic farms will face with respect to Medicaid funding.)

As yet another example, in the state of Ohio, funding is potentially available to Hopewell through the Alcohol, Drug Addiction and Mental Health Services Board of Cuyahoga County, which receives its funding from both the state and local governments. Under Ohio law, the Alcohol, Drug Addiction and Mental Health Services Board of Cuyahoga County is one of 50 such boards in the state charged with the responsibility of providing mental health and addiction treatment services to individuals within its jurisdiction through the awarding of contracts to local provider agencies (Alcohol Drug Addiction and Mental Health Services Board of Cuyahoga County, n.d.). However, whether Hopewell would actually receive funding at a given point in time would depend on the extent to which any funding is available, the funding priorities of the board and its funders at a particular point in time, and Hopewell's ability to compete successfully against other mental health treatment providers in securing that funding.

Due to the relative scarcity of government funding, US-based therapeutic farms must often rely on private funding to remain sustainable. This funding may be derived through various mechanisms including annual large-scale fundraising events; donations from private individuals and foundations; agency and foundation grants; bequests; and fees paid by residents, their family members, and/or insurance companies for the costs associated with their care. The costs of care for an individual vary widely depending upon the amount of funding otherwise available to a therapeutic farm to sustain its programming, relative to the costs of that programming and the number and cost of add-on services requested by a resident, such as additional transportation services and additional psychotherapy sessions beyond what is usually provided.

Challenges Facing Therapeutic Farms

Therapeutic farms are facing significant financial challenges. Residential care is relatively expensive and relatively few individuals are able to afford the cost of care from their own funds. Often health-care insurance, the primary mechanism for the payment of health care in the United States, frequently imposes limits on individuals' length of stay in a facility or the amount of coverage available for residential mental health care. Individuals who lack adequate private funding for the cost of their care and whose insurance places limitations on the length of stay, the cost of residential care, or the type of services that insurance will pay for during their residential care may be unable to access needed services provided by the therapeutic farm, absent the availability of a scholarship or donated funds to cover the cost of care. This has implications not only for individuals who may be in need of such services but for the financial survival of the farms themselves.

The financial challenges are further amplified by the regulatory patchwork that governs the farms' existence. These regulations relate to required staffing, required reporting about the residents' care, and/or the maintenance of the farm, the farm's employees, and the animals and produce. Many of these regulations were promulgated in an effort to safeguard the well-being of the consumers of mental health services and/or to ensure that the care furnished by all facilities adhered to at least a minimum standard. However, significant staff time is often required to comply with the regulations and to document that compliance. A failure to comply can lead to expensive fines and serious sanctions.

Additionally, the complex regulations often necessitate legal consultation, which is also expensive. Such consultation and representation may be necessary to address state investigations of incidents at the facility, such as a resident's death; to respond to state audit findings; and/or in the event of a lawsuit by a resident who believes that he or she was not treated well or by the family of a resident.

Although the leadership of the existing therapeutic farms is generally close to the age of retirement, generally about 65–70 years of age, there has been little succession planning. Few of the existing therapeutic farm communities are able to identify an already-trained, competent successor should the existing director retire, resign, become ill, or die. This is equally true of many of the members of the farms' governing boards.

References

Alcohol Drug Addiction and Mental Health Services Board of Cuyahoga County. (n.d.). *ADAMHS Board of Cuyahoga County*. http://www.adamhscc.org/. Accessed 4 May 2015.

Barnes, M. (2008). Care, deliberation and social justice. In J. Dessein (Ed.), *Farming for health. Proceedings of the community of practice farming for health, November 2007, Ghent* (pp. 27–35). Merelbeke: ILVO.

Centers for Medicare and Medicaid Services. (n.d.). *Medicare.gov*. http://www.medicare.gov/your-medicare-costs/help-paying-costs/medicaid/medicaid.html. Accessed 4 May 2015.

de Krom, M. P. M. M., & Dessein, J. (2013). Multifunctionality and care farming: Contested discourses and practices in Flanders. *Netherlands Journal of Agricultural Science—Wageningen Journal of Life Sciences, 64*(65), 17–24.

De Bruin, S. R., Oosting, S. J., Tobi, H., Blauw, Y. H., Schols, J. M. G. A., & de Groot, C. P. G. M. (2010). Day care at Greek care farms: A novel way to stimulate dietry intake of community-dwelling older people with demetia? *Journal of Nutrition, Health, and Aging, 14*(5), 352–357.

Dessein, J. (2008). Introduction. In J. Dessein (Ed.), *Farming for health. Proceedings of the Community of Practice Farming for Health, November 2007, Ghent* (pp. 1–11). Merelbeke: ILVO.

Dessein, J., Bock, B. B., & de Krom, M. P. M. M. (2013). Investigating the limits of multifunctional agriculture as the dominant frame for Green Care in agriculture in Flanders and the Netherlands. *Journal of Rural Studies, 32*, 50–59.

De Vries, S. (2004). Contributions of natural elements and areas residential environments to human health and well-being. In J. Hassink & M. van Dijk (Eds.), *Farming for health: Green care Farming across Europe and the United States of America* (pp. 21–30). Dordrecht: Springer.

Di Iacovo, F., & O'Connor, D. (Eds.). (2009). *Supporting policies for social farming in Europe. Progressing Multifunctionality in responsive rural areas ("SoFar Project")*. Florence: Agenzia Regionale per lo Sviluppo e l'Innovazione nel Settore Agricolo-forestale (ARSIA).

Friedel, S., Mathijs, E., & Van Molle, L. (2010). *Policy and frames regarding social farming and green care in Flanders and at the EU level: Working paper 2010/103*. Leuven, Belgium: Centre for Agricultural and Food Economics, Katholieke Universiteit Leuven. http://www.biw.kuleuven.be/aee/clo/wp/friedel2010.pdf. Accessed 25 Mar 2015.

Gould Farm. (n.d.). *About us*. http://www.gouldfarm.org/about-us. Accessed 1 Apr 2015.

Hassink, J., Hulsink, W., & Grin, J. (2012). Care farms in the Netherlands: An underexplored example of multifunctional agriculture—Toward an empirically grounded, organization-theory-based typology. *Rural Sociology, 77*(4), 569–600.

Hassink, J., Zwartbol, C., Agricola, H. J., Elings, M., & Thissen, J. T. N. M. (2007). Current status and potential of care farms in the Netherlands. *Netherlands Journal of Agricultural Science—Wageningen Journal of Life Sciences, 55*(1), 21–35.

Haubenhofer, D. K., Elings, M., Hassink, J., & Hine, R. E. (2010). The development of green care in western European countries. *Explore—The Journal of Science and Healing, 6*(2), 106–111.

Hine, R., Peacock, J., & Pretty, J. (2008). *Care farming in the UK: Evidence and opportunities, Report for the National Care Farming Initiative (UK)*. Essex: University of Essex.

Iancu, S. C., Zweekhorts, M. B. M., Veltman, D. J., van Balkom, A. J. L. M., & Bunders, J. F. G. (2015). Outsourcing mental health care services? The practice and potential of community-based farms in psychiatric rehabilitation. *Community Mental Health Journal, 51*, 175–184.

Internal Revenue Code § 501(c)(3) (2014).

McGloin, A. (2009). A journey through social farming in Europe: The case studies, Ireland. In F. Di Iacovo & D. O'Connor (Eds.), *Supporting policies for social farming in Europe. Progressing multifunctionality in responsive rural areas ("SoFar Project")* (pp. 98–109). Florence: ARSIA.

Ohio Secretary of State, Business Services Division. (n.d.). *Your guide to starting a nonprofit in Ohio*. Columbus, OH: Author. http://www.sos.state.oh.us/sos/upload/publications/busserv/Nonprofit.pdf. Accessed 2 Apr 2015.

Roest, A. E., Oosting, S. J., Ferwerda-van Zonneveld, R. T., & Caron-Flinterman, F. (2010). Regional platforms for green care farming in the Netherlands. In *Ninth European IFSA symposium, Vienna*, July 4–7. http://ifsa.boku.ac.at/cms/fileadmin/Proceeding2010/2010_WS1.2_Roest.pdf. Accessed 27 Mar 2015.

Spring Lake Ranch Therapeutic Community. (n.d.). *The history of Spring Lake Ranch Therapeutic Community*. http://www.springlakeranch.org/about/history. Accessed 2 Apr 2015.

United States Department of the Treasury, Internal Revenue Service. (2014). *Applying for 501(c)(3) tax-exempt status*. (Pub. 4220, rev. 7–2014). Washington, DC: Author. http://www.irs.gov/pub/irs-pdf/p4220.pdf. Accessed 2 Apr 2015.

United States Department of the Treasury, Internal Revenue Service. (2015). *Tax-exempt status for your organization*. (Pub. 557 rev. Feb. 2015). Washington, DC: Author. http://www.irs.gov/pub/irs-pdf/p557.pdf. Accessed 2 Apr 2015.

Van Schaik, J. (1997). *Where agriculture and care meet. Stock taking of practical experiences on care farms*. Vorden: Stichting Omslag. Cited in Roest, A. E., Oosting, S. J., Ferwerda-van Zonneveld, R. T., & Caron-Flinterman, F. (2010). Regional platforms for green care farming in the Netherlands. In *Ninth European IFSA symposium, Vienna*, July 4–7. http://ifsa.boku.ac.at/cms/fileadmin/Proceeding2010/2010_WS1.2_Roest.pdf. Accessed 27 Mar 2015.

Vik, J., & Farstad, M. (2009). Green care governance: Between market, policy and intersecting social worlds. *Journal of Health Organization and Management, 23*(5), 539–553.

Chapter 4
Hopewell Therapeutic Farm Community

Hopewell Therapeutic Farm Community: Overview

Hopewell Therapeutic Farm Community was established as a private charitable nonprofit corporation in Mesopotamia, Ohio, on July 4, 1993, under the name of Hopewell Inn. The facility was the brainchild of Clara Rankin, whose son had been diagnosed with schizophrenia. He had spent time at the United States' first therapeutic farm community for adults with severe mental illness, Gould Farm in Massachusetts. Mrs. Rankin, believing that additional such accommodations were necessary to provide alternatives for the treatment of mental illness, set about raising funds and securing a location.

Hopewell is located in Trumbull County, Ohio, a largely rural area that is home to the fourth largest Old Order Amish population in the world. Hopewell consists of more than 300 acres of land that encompass woods, trails, a lake, hills, a main building that includes a common dining rooms and clinical offices, a music conservatory, four residence buildings for clients, four residence buildings for staff, six barns, and a country store. An additional building, referred to as the depot, contains the Hopewell library on its first floor and provides apartment housing for families and visitors on its second floor. On average, Hopewell provides services to 30–35 residential clients. Past clients can continue to access transitional services through Hopewell's Club Hope. Additionally, Hopewell's recently acquired family home in the nearby town of Mesopotamia provides housing for five clients who are transitioning from residence at Hopewell to more independent living situations (Fig. 4.1).

© Springer International Publishing Switzerland 2016
S. Loue, *Therapeutic Farms*, SpringerBriefs in Social Work,
DOI 10.1007/978-3-319-13539-7_4

Fig. 4.1 Hopewell residents' housing

A typical Hopewell day begins with a morning meeting, where announcements and good news are shared. This is followed by work crews and school, a noon hour lunch when everyone eats together, and then afternoon work crews and therapeutic groups. Later in the day, there are meetings of the Resident Community Council and either cottage meetings of the residents in their building or an All-Community Meeting. Everyone dines together for the evening meal.

Hopewell provides individualized treatment within a community setting. Residents hold each other accountable for socially acceptable behavior that reflects the four core values of self-discipline, moderation, integrity, and justice. Each of these terms is defined so that the goals are understood and shared across the Hopewell community. Self-discipline encompasses honesty, responsibility, radical acceptance, attendance, timeliness, and good hygiene. Moderation is required in thought, word, and deed and evidences self-restraint and willpower. Integrity refers to standing up for one's own beliefs and demonstrating courage and strength of mind. Justice requires that others be treated fairly and that actions and word conform to truth, fact, and reason. Each resident's progress is tracked on a daily basis, and as progress is made, the resident's privileges are gradually increased from admission through three interim phases to the transition planning phase that focuses on preparation for the resident's return to the community. Figures 4.2 and 4.3 explain the point system utilized to achieve various phases and provide an example of a typical week of activities.

POINTS–Revised February 2, 2015

TIME	Monday		Tuesday		Wednesday		Thursday		Friday		Total
9:00 – 10:00	Work Crew	2	Work Crew	2	Work Crew	2	Work Crew	2	Work Crew	2	
10:00 – 11:00	Work Crew	2	Work Crew	2	Work Crew	2	Work Crew	2	Work Crew	2	
11:30 – 12:00	Morning Meeting	1	Morning Meeting	1	Morning Meeting	1	Morning Meeting	1	Morning Meeting	1	
12:30 – 5	Outdoor WC Reading Mental Health Life Skills	1	Outdoor WC Education Spirituality Horsemanship Process Group	1	Weaving Nature Group Recovery Group	1	Outdoor WC Art Process Group	1	Outdoor WC Interpersonal Skills EAL Creative Writing	1	
3:30	Cottage or Community Meeting	1	Mindfulness								
	Med Fill	1									
TOTAL		8		6		6		6		6	32

1. There are 32 possible points during the regular weekday schedule.
2. There are many opportunities to earn extra such as weekend Farm Crew (2 points for morning crew, 1 for afternoon), and other activities assigned/negotiated with staff such as snow shoveling or extra cleaning.
3. You need 32 points to get 100%
4. You cannot get over 100% If you earn more than 32 points, that is wonderful, you are really taking full advantage of all that Hopewell has to offer, great job !
5. Phase 3 – You need at least 29 points to maintain above a 90% in order to earn or not risk losing Phase 3
6. Phase 2 – You need at least 26 points to maintain above a 80% in order to earn or not risk losing Phase 2
7. Phase 1 – You need at least 16 points to maintain 50% in order to earn Phase 1

Fig. 4.2 Point system to attain various phase statuses

SCHEDULE – revised February 2, 2015

Effective 2/2/15	MONDAY	TUESDAY	WEDNESDAY	THURSDAY	FRIDAY	SATURDAY	SUNDAY
7:30am-8:15am	Breakfast	Breakfast	Breakfast	Breakfast	Breakfast	Breakfast	Breakfast
8:30am	Meditation, Med Fill, Exercise, Jam	Meditation, Med Fill, Exercise, Jam	Meditation, Med Fill, Exercise, Jam	Meditation, Med Fill, Exercise, Jam	Meditation, Med Fill, Exercise, Jam		
9am-11:00am	Work Crews	Work Crews	Work Crews	Work Crews	Work Crew	10am Farm Crew	10am Farm Crew
11:30 -12	Morning Meeting (Dining Room)	Morning Meeting (Dining Room)	Morning Meeting (Dining Room)	Morning Meeting (Dining Room)	Morning Meeting (Dining Room)	Free Time	Worship Trips
12pm-12:30pm	Lunch	Lunch	Lunch	Lunch	Lunch	Lunch	11:30 Brunch
1:00pm 2:00pm	Outdoor Work Reading (Library) Mental Health Education (Conservatory)	Outdoor Work Education (Crows Nest) Spirituality (Dining Room) Horsemanship (South Farm)	Fiber Arts (Dining Room) 1-2:30 Nature Group (Announced)	Outdoor Work Art (Art Room) (Dining Room) (Pottery Studio)	12:30-1:30pm Room Cleaning 1:30 - 2:30pm Outdoor Work Equine Assisted Learning (South Farm closed) 1:45 - 2:30pm Interpersonal Relationship Skills (Dining Room)	Free Time Recreational &social trips	Free Time Recreational &social trips
2:00pm 3:00 pm		2:00-3:00pm Outdoor Work	2:15-3:15pm Recovery Group (Closed Art Room)	2:00-3:00pm Outdoor Work			
2:30pm 3:30pm	2:30 - 3:15pm Life Skills (Dining Room - Inn)	Process Group (Cottage 2 LR) Hopewell Store		Process Group (Cottage 2 LR) Hopewell Store			
3:00pm 5:00pm	3:30pm Cottage or Community Meeting 4:00 - 4:30pm Resident Council (when announced) (Dining room)	3:00pm - 4:00pm Mindfulness & Meditation (Conservatory - closed)			3:00-4:00pm Writing Group (Cottage 2)	4pm Farm Crew	4pm Farm Crew
5:30pm - 6:00pm	Dinner	Dinner	Dinner	Dinner	Dinner	Dinner	Dinner
Evenings	Recreation/Outings Computer Time	Recreation/Outings Computer Time	Recreation/Outings Computer Time	Recreation/Outings Computer Time	Recreation/Outings Computer Time	Recreation Computer Time	Recreation Computer Time
9:30pm	Quiet Time	Quiet Time	Quiet Time	Quiet Time	Quiet Time (10:30)	Quiet Time (10:30)	Quiet Time

Fig. 4.3 Typical schedule of daily activities

As indicated in Fig. 4.4, achievement of a phase enables the individual to participate in additional Hopewell activities, including outings to the YMCA, bowling,
and local stores for shopping.

 # June 2015

Sun	Mon	Tue	Wed	Thu	Fri	Sat
	1	2 6:15 pm Library Ph.I/II/III/ES	3 6:15 pm YMCA Ph.I/II/III/ES	4 12:45/6:15 pm Wal-Mart Ph. II/III/ES	5	6 12:30 pm Bowling Ph.I/II/III/ES
7 12:30 pm YMCA Ph.I/II/III/ES	8	9 6:15 pm Library Ph.I/II/III/ES	10 6:15 pm YMCA Ph.I/II/III/ES	11 6:15pm Movie Trip ph.III/ES	12	13 12:30 pm Bowling Ph.I/II/III/ES
14 12:30 pm YMCA Ph.I/II/III/ES	15 6:15 pm Commons Ph.II/III/ES	16 6:15 pm Library Ph.I/II/III/ES	17 6:15 pm YMCA Ph.I/II/III/ES	18 12:45/6:15 pm Wal-Mart Ph. II/III/ES	19	20 12:30 pm Bowling Ph.I/II/III/ES
21 12:30 pm YMCA Ph.I/II/III/ES	22 6:15 pm Dollar Store Ph. I/II/III/ES	23 6:15 pm Library Ph.I/II/III/ES	24 6:15 pm YMCA Ph.I/II/III/ES	25	26	27 12:30 pm Bowling Ph.I/II/III/ES
28 12:30 pm YMCA Ph.I/II/III/ES	29 6:15 pm Wal-Mart Trip Ph. I	30 6:15 pm Library Ph.I/II/III/ES Mall Trip 8:30 am Ph.II/III/ES				
					Date for Super Trip/lunch out will be announced by Resident Coun-	4:15 pm Haircuts @ Hopewell $10/$12 Haircuts $5 Bang/Beard Trim

Fig. 4.4 Schedule of Hopewell outings, by day and phase status, June 2015

Residents who have achieved phase three status are eligible to participate in a community council. The council is responsible for problem-solving in the case of disputes between residents; for considering and imposing trip restrictions when a resident has demonstrated disruptive, unsafe, or disrespectful behavior; and for the evaluation, approval, and denial of "promotion" to the next phase. The final decision regarding promotion through the various phases rests with each resident's counselor/therapist.

The sense of community and the individualized care that Hopewell provides are of critical importance to many residents. One resident, discharged from Hopewell in 2013, remarked, "Hopewell offers an individualistic treatment for each resident, which is very much different from the stereotypical mental health treatment program." The following comments of residents at the time of their discharge during the period from 2012 through 2014 are representative of past Hopewell clients regarding their experience of community at Hopewell:

> Hopewell saved my life. I never thought I would be happy again. I will always feel indebted to them.

> Thank you for this amazing once in a lifetime chance to find myself. With this community's help I was able to work on things I never thought possible.

[I was helped by] the sense of community and how each member of the community is important for the success of Hopewell … I am more than satisfied. Hopewell helped me find the true me and become more in tune with my mental and physical importance.

Being in a small society and interacting with other people so I could get used to it in the real world.

Hopewell pretty much gave me a life that I have always wanted to have. It allowed me to see who I was—what I liked about me and what I wanted to change. I don't think I'll ever have another opportunity like Hopewell again.

I have been in need of recovery for many years. A residential farm community like Hopewell provided the safety and care I think I have been looking for. I developed a sense of hope here—one I didn't have coming in.

This place has shaken up my world and changed my life. The sense of belonging, community and the wellness of others has [sic] helped me heal.

I feel like my load has lifted and I have a satisfying plan for tomorrow.

Programming

Work

Hopewell maintains a 5-day-a-week work program that focuses on helping clients/residents develop life preparedness skills that they will need in order to obtain and sustain meaningful employment. Although the specific form of work offered depends upon each individual's readiness and level of ability, all work programming emphasizes development and maintenance of the following skills:

- Being ready and able to work, e.g., timeliness, appropriate dress, and willingness to work
- Acceptance of guidance, instruction, and direction
- Completion of tasks
- Cooperation with team leaders
- Cooperation with peers

Work is intended to serve multiple purposes. First, it provides residents with a sense of structure and balance. Residents witness the cycle of the seasons and the cycle of the days. Work chores must be performed at designated times of the day and the year. Spring brings the planting season and summer the gathering of the crops. Autumn brings maple syrup season, when the sap must be drained from the trees. The animals must be fed each morning, no matter what the time of year or the temperature outside.

Work crews operate for 2 h each day. Residents can choose to participate in any of the following work crews.

- *Kitchen Crew* members help with dishwashing, fill and restock items for the salad bar, and set up the dining room for lunch.

Fig. 4.5 Hopewell cows

- *Farm Crew* members are responsible for the care of all Hopewell animals; the removal of collected trash from Hopewell buildings; and the collection, disinfection, and storage of fresh eggs. Residents can elect to participate on Farm Crew on Saturdays, as well as during the week, in order to ensure that the farm animals receive their needed care and the barn chores are done (Fig. 4.5).
- *Housekeeping Crew* participants vacuum or mop floors, disinfect door knobs, clean chairs, wash windows, water indoor plants, collect garbage and recycling, clean bathrooms, restock bathroom supplies, clean closets, and collect and deliver laundry items.
- *Garden/Art Crew* members, who work in the garden between February and October, are responsible for preparing plants, weeding, harvesting produce for the kitchen and for sale at the market, decorating the grounds and buildings with seasonal displays, and assisting with leaf and snow removal during autumn and winter.
- *Maintenance Crew* members meet each day to assess and then address Hopewell's maintenance needs, which may include snow and leaf removal, flower bed upkeep, repair of driveway potholes, power washing and cleaning buildings, the relocation of furniture and/or residents, grass cutting, and weed whacking.

Residents can also participate in activities related to the Hopewell Store and the Farm and Craft Market. The store, which is open twice each week, sells food and drink items to residents and staff. The store, in particular, has helped residents learn new skills and maintain existing ones. These include how to operate a cash register, greet people, manage a store inventory, and maintain a financial ledger. Individuals who have participated in operating the store functions have reported that these activities have helped to improve their self-esteem and have made them feel valued. The Farm and Craft Market sells art objects made by the residents, such as wooden cutting boards, and produce from the Hopewell garden.

Many residents have remarked on the positive effects of a structured work day and work itself. The following provides a sample of comments from clients who resided at Hopewell during the years 2012 through 2014:

Work crews gave me a sense of purpose.

Because I see working is a normal behavior and doing it makes me happy.

The structure [helped me most]. Being able to have a busy day without having to make 100 choices throughout the day.

Morning work crew gave me a sense of belonging throughout the day's activities.

Working in the morning helped me get in a routine of getting things done first thing.

Housekeeping helped me relearn organization skills.

Working on maintenance crew gives me a sense of purpose and work ethic while I work on my communication skills.

[Work crew] helped me with distress; there have been times when I am consumed by stress-ful thinking.

Residents have often been particularly positive about their work experiences with the animals. One resident, who left Hopewell in 2014, said "I really connected with the horses." Another, who left in 2012, observed, "Walking in the woods and attending to the animals helped me be at peace with the natural world."

Hopewell is currently developing a formal on-the-job training program to provide residents with opportunities to learn specific skills that can then be translated directly into employment situations. As residents' skill level increases, they will be transitioned from unpaid on-the-job training slots to paid employment on the farm. It is anticipated that many of the tasks for which they will be paid at Hopewell will be similar to those that might encounter once they leave. As an example, individuals who wish to continue living in a more rural area will be trained to care for farm animals. As they develop greater ease and expertise, they will be compensated for these tasks, which are the same as those that they would perform on many of the farms in the areas near Hopewell.

Music and Art

Hopewell maintains a vibrant art and music program for residents. Creative art activities are available on an almost daily basis and Hopewell maintains a dedicated art room for these activities. Hopewell is in the process of adding a part-time art therapist to its staff. The music education program, which includes individual lessons on a variety of musical instruments, is overseen by a staff member who was once a long-time music teacher. A number of Hopewell residents have formed a band and periodically offer performances both at Hopewell and in the larger community (Fig. 4.6).

Fig. 4.6 Hopewell conservatory

Nature Programming

Hopewell's nature programming is based on the idea that nature contains teaching and laws that are both valuable in clients' lives and that can promote mental health recovery. These teachings can be learned both didactically and experientially (Fig. 4.7).

The program seeks to enhance clients' mental health, develop clients' sense of connection with the natural world, increase clients' sense of identity with a larger ideal, enhance the social connections within and between groups, increase clients' awareness of the importance of environmental stewardship, provide clients with a new sense of boundaries and understanding of the consequences of their behavior, improve clients' physical health through engagement in physical activity such as hiking, and help clients' develop new knowledge and skills and a sense of competency related to outdoor work and nature (Ruch and Carlton, n.d.).

There are two nature groups each week. On one day, the group walks on the trails, with psychoeducation linked to didactic education about ecology, ornithology, geology, and forestry. The second group is conducted inside to facilitate processing of the previous day's experiences and to accommodate those who may be unable to go outside. Table 4.1 provides an example of both a didactic lesson and an experiential activity (Ruch and Carlton, n.d.).

Fig. 4.7 Hopewell trail

Table 4.1 Example of nature group didactic and experiential lessons

Didactic Sample
Theme: Transition to spring, part 4—orienteering and goal setting
Activities: Teach basic group orienteering skills such as shooting an azimuth, getting bearings, and utilizing different aspects of the compass. Encourage the concept of orienteering as an analogy for making goals
Interventions: Encourage group members to participate in experiential activity of using compass to connect to the natural world. Facilitate exploration of goal setting and getting an ideal for goal setting
Helpful knowledge/materials for leader:
Biological/geographical knowledge: Knowledge of geography and landscape; knowledge of how to use a compass
Psychotherapy knowledge: Knowledge of setting and meeting goals
Materials: Compass; open field/space to shoot azimuths
Experiential Example
Goal: Experiencing a field to benefit mood
Activities: Have group explore a field and observe all the plant and animal life and take pictures. Encourage members to experience different aspects of the field such as bird sounds and beauty of flowers and smelling flowers and touching plants. Ask members to note any positive changes in mood and to give one-word phrase denoting positive mood change as to how nature makes them feel (such as "peaceful")
Interventions: Support group members in their exploration of their feelings with nature experiences
Helpful knowledge/materials for the leader:
Biological knowledge: Knowledge of spring plant/bird/animal identification, ecology, and ornithology
Psychotherapy knowledge: Knowledge of how to encourage group members to engage in progressive relaxation exercises
Materials: Hiking boots, camera, bug repellant

Hopewell's nature-based programming includes a sensory trail. The sensory trail consists of 12 posts placed along the various trails. Each post bears a letter, and each letter corresponds to a letter in the picture hunting activity folder that is provided to residents. Each letter in the activity folder contains a lesson involving an experiential activity at that location post (Figs. 4.8 and 4.9a, b).

Fig. 4.8 A Hopewell trail

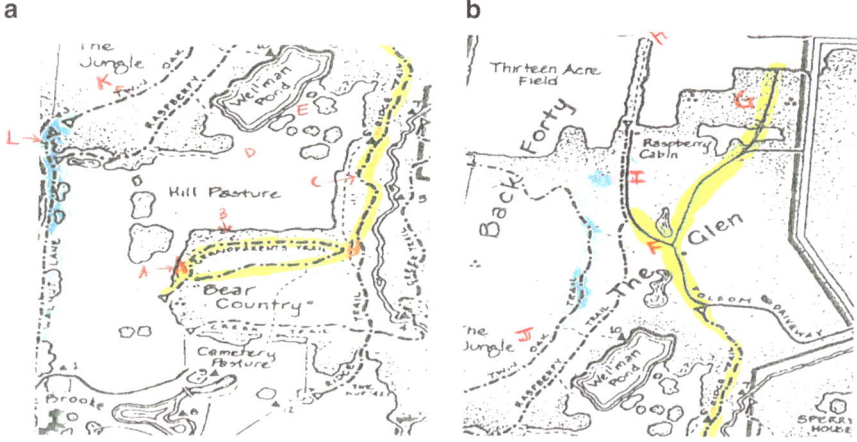

Fig. 4.9 (**a**). Sensory trail overview. (**b**) Sensory trail detail

Table 4.2 Example of location post activity on Hopewell's sensory trail

Location E. Wellman Pond

Wellman Pond is a nice spot to relax and watch the natural life of the pond go about its business. From here you can probably see birds of different types, different plants, and also maybe some fish or turtles below the surface of the water.

Experiential activity:
Please take a moment to sit down by the location post and see what birds, fish, insects, and plants you can see by the pond. Please close your eyes and listen to any possible sounds of life that you can hear in the water around Wellman Pond.

Question for reflection:
How do these sights and sounds affect your mood? Does the water have a calming effect on your mood?

The sensory trail is intended to help residents relax and meditate. They can walk the sensory trail sequentially in order of the alphabet or in any random order. Completion of all of the activities at all of the posts requires just over 1 h. Table 4.2 provides an example of one of the location post activities.

One resident offered the following comment on the nature-based program: *"It was good to get in touch with nature to watch nature and become closer to it."* Others have stated that they appreciated *"walking in the woods," "the open air … being able to walk out in the woods or fish," "the sense of connectedness to nature,"* and *"walks—they cleared my mind."*

Medication Management, Counseling, and Therapy

Hopewell provides pharmacologic management and a consulting psychiatrist visits on a regular basis. Hopewell holds and protects clients' medication and assists clients with self-administration of medications as clinically appropriate. Transportation is provided to routine medical appointments.

Numerous residents have indicated at discharge that Hopewell helped them both realize the importance of their medications and develop a routine for taking them. In response to the question, "What part of Hopewell helped you most in your recovery," some clients indicated:

getting put on the right meds for me

I got the right diagnosis and medications and I was able to grow emotionally and spiritually in a very therapeutic environment.

Counseling and therapy are provided to clients, on an individual and/or group basis, either directly or by way of referral, to address a wide range of issues, as needed. These include:

- Substance abuse issues
- Mental health issues
- Family issues

Fig. 4.10 Horses in the equine-assisted program

- Cognitive functioning
- Emotional functioning
- Improvement of coping abilities
- Social functioning
- Time management
- Relapse prevention
- Community living skills

Clinical groups include dialectical behavior therapy (DBT), the equine-assisted program (EAP), horsemanship, independent living skills, interpersonal relationship skills, mental health, mood management, nature-based group, resident council, spirituality, and posttraumatic stress disorder (Fig. 4.10).

Notably, Hopewell is the only one of the US-based therapeutic farms that is accredited by the Commission on Accreditation of Rehabilitation Facilities (CARF) for the provision of mental health services. CARF is a nonprofit organization dedicated to the promotion of "quality, value, and optimal outcomes of services through a consultative process and continuous improvement services that center on enhancing the lives of persons served" (CARF International, 2015a). Accreditation is conferred only after an internal examination of the program and its business practices, an on-site survey conducted by a CARF-selected team of experts, and provider adherence to CARF standards and dedication to and engagement in continuous improvement (CARF International, 2015b).

Many clients have offered comments, such as the following ones, regarding the helpfulness of the therapeutic services that they received:

I was able to get off all psychotropic medications and use tools like CBT that I have much more faith in given my personal history with mental health treatment, as well as CBT therapy being more appropriate for my particular psychological struggle.

The one-on-one sessions helped me most because it made me feel like I could open up more.

Hopewell gave me the tools to keep my temper under control and be more tolerant of other people.

It [Hopewell] helped to learn about my mental illness and how to manage it.

One-on-one therapy with my clinician helped me understand my diagnosis and provided me with concrete methods to deal with my symptoms.

My clinician [name] helped me the most because we were able to communicate very effectively.

Meeting with [my therapist] helped me become aware of the cause of my depression.

[Therapy] groups give me a sense of belonging.

The [mental health] group helps me arm myself with education to use in my battles with mental illness for a most effective and positive tool in my recovery.

The mood management group helps you figure out what you are feeing and figure other ways to help you with your moods.

Additional Activities

Residents have the option of participating in various groups that are geared to specific interests. These include a sports activities group that provides opportunities to go bowling or to a health club; a reading group; a weaving and fine arts group; a creative expression group; an education group that assists residents with the development of reading and writing skills; and a life skills management group that focuses on budgeting, financial management, and relocation services.

Program Effectiveness

Hopewell dedicates significant resources to evaluate on an ongoing basis the effectiveness of the programming that it provides and maintains a rigorous quality improvement program. Program evaluation utilizes both quantitative methodologies, such as comparing scores on various measures at admission with scores on the same instruments at the time of discharge, and qualitative methodologies, such as resident questionnaires and interviews that ask for residents' opinions about the services that they received. As an example, Hopewell utilizes residents' scores on the Global Assessment of Functioning (GAF) at the time of admission and discharge to assess whether the program has helped to improve residents' ability to function. (Although the GAF is no longer recognized in the most recent *Diagnostic and Statistical Manual of Mental Disorders (DSM-V)* (American Psychiatric Association, 2013), published on May 18, 2013, Hopewell has found and continues to find it invaluable in tracking residents' progress over time.) As Fig. 4.11 illustrates, a recent assessment of the residents' ability to function shows a consistent increase between the time of admission and discharge for the period from June 2006 through December 2014. These results suggest that Hopewell programming is, indeed, effective at helping residents function more independently.

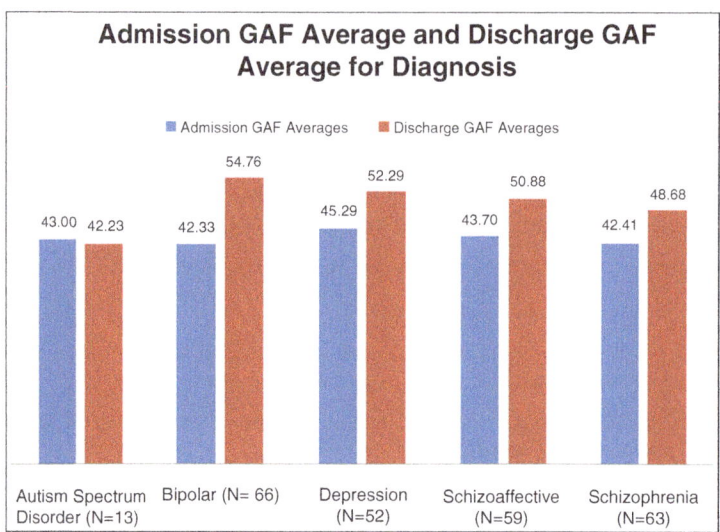

Key: Global Assessment of Functioning Scale

91 - 100 No symptoms. Superior functioning in a wide range of activities, life's problems never seem to get out of hand, is sought out by others because of his or her many positive qualities.

81 - 90 Absent or minimal symptoms (e.g., mild anxiety before an exam), good functioning in all areas, interested and involved in a wide range of activities, socially effective, generally satisfied with life, no more than everyday problems or concerns (e.g., an occasional argument with family members).

71 - 80 If symptoms are present, they are transient and expectable reactions to psychosocial stressors (e.g., difficulty concentrating after family argument); no more than slight impairment in social, occupational or school functioning (e.g., temporarily falling behind in schoolwork).

61 - 70 Some mild symptoms (e.g., depressed mood and mild insomnia) *or* some difficulty in social, occupational or school functioning (e.g., occasional truancy or theft within the household), but generally functioning pretty well, has some meaningful interpersonal relationships.

51 - 60 Moderate symptoms (e.g., flat affect and circumstantial speech, occasional panic attacks) *or* moderate difficulty in social, occupational or school functioning (e.g., few friends, conflicts with peers or co-workers).

41 - 50 Serious symptoms (e.g., suicidal ideation, severe obsessional rituals, frequent shoplifting) *or* any serious impairment in social, occupational or school functioning (e.g., no friends, unable to keep a job, cannot work).

31 - 40 Some impairment in reality testing or communication (e.g., speech is at times illogical, obscure or irrelevant) *or* major impairment in several areas, such as work or school, family relations, judgment, thinking or mood (e.g., depressed adult avoids friends, neglects family and is unable to work; child frequently beats up younger children, is defiant at home and is failing at school).

21 - 30 Behavior is considerably influenced by delusions or hallucinations *or* serious impairment in communication or judgment (e.g., sometimes incoherent, acts grossly inappropriately, suicidal preoccupation) *or* inability to function in almost all areas (e.g., stays in bed all day, no job, home or friends)

11 - 20 Some danger of hurting self or others (e.g., suicide attempts without clear expectation of death; frequently violent; manic excitement) *or* occasionally fails to maintain minimal personal hygiene (e.g., smears feces) *or* gross impairment in communication (e.g., largely incoherent or mute).

1 - 10 Persistent danger of severely hurting self or others (e.g., recurrent violence) *or* persistent inability to maintain minimal personal hygiene *or* serious suicidal act with clear expectation of death.

0 Inadequate information

Fig. 4.11 Average GAF (Global Assessment of Functioning) scores at admission and discharge, by diagnosis, June 2006–December 2014

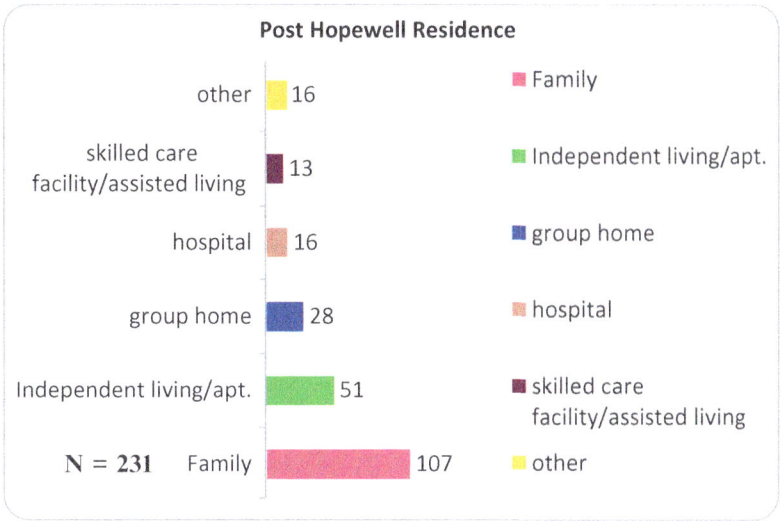

Fig. 4.12 Post-Hopewell residence June 2006–December 2014 ($n = 231$)

A review of post-Hopewell residential arrangements for departing residents further suggests that the program aids clients in becoming more self-sufficient. A total of 186 of 231 clients (80.5 %) discharged between June 2006 and December 2014 have been able to reestablish residence with their family members (spouses, significant others, siblings, or parents) or arrange for residence in group homes or independent living situations, such as their own apartments. See Fig. 4.12.

Visioning the Future

Hopewell undertook an in-depth strategic planning process in 2013 designed to serve as a blueprint for the upcoming 3 years. The process focused on five specific areas: existing services, research and new service development, marketing and sales, fund development, and human resources. Staff and board members were involved in all phases of the assessment and planning process. Key goals in each of these domains are noted in Table 4.3.

The focus on research, with particular emphasis on identifying the level of effectiveness associated with each component of the program, is especially notable. Current research focuses on assessing past residents' satisfaction with the programming available at Hopewell, the identification of gaps in services, and the identification of program components that hold a greater likelihood of yielding positive outcomes for specific groups of clients, e.g., those experiencing depression or those in a certain age range. Research findings will enable Hopewell administrators to evaluate the need for various programs, allow clinicians to better target specific

Table 4.3 Goals of Hopewell's strategic action plan for the years 2013–2016

Domain	Goal	
Existing services	Quality of service	• Establish a formalized long-term program licensed by the Ohio Department of Mental Health • Collect 2 years of outcome data on long-term and transitional programs
	Staff development	Have 100 % of staff complete training in at least one evidence-based program in accordance with the training plan
	Cost of service	• Unbundle fees; bill for assessments • Establish a process that effectively integrates billing and service provision
	Program	• Complete an assessment of unbundled services and take appropriate action based on the effectiveness of such services • Institute evidence-based practice and ensure that all modifications and enhancement are channeled through this process
	Efficiency	• Combine the intake and diagnostic assessments
Research and new service development	Research	• Publish professional articles and present at professional conferences • Complete grant proposal to support additional research • Assess outcomes of programming
	New services	• Review literature relating to efficacy of new services, e.g., vocational services, aftercare, and alternative therapies, and assess feasibility of their implementation
Marketing and sales	• Expand dissemination of information about Hopewell to potentially interested audiences • Advertise services through social media • Revamp logo and advisability of name change • Upgrade website • Provide training to the marketing director • Engage external marketing firm	
Fund development	• Increase foundation and endowment support • Coordinate and organize fundraising events, e.g., annual fund campaign • Increase focus on gift planning • Develop new fundraising events and opportunities • Increase board member training and involvement in various aspects of fundraising • Provide ongoing communication to major donors	
Human resources	Staff development and satisfaction	• Conduct 360° evaluations • Create succession ladder • Implement staff development plans • Develop Personal Professional Development Plans • Develop improved means to communicate within a 24-h day organization
	Volunteer program	• Identify potential volunteer opportunities • Formalize and emphasize the volunteer program • Cultivate relationships with additional sources of volunteers
	Board composition and engagement	• Recruit and select board members based on institutional needs • Ensure that all board meetings are engaging and purposeful

types of interventions to those clients who are more likely to derive benefit from them, and provide consumers with more specific information related to the effectiveness of the various program components. Ultimately, this focused attention on evaluating outcomes will ensure that, to as great a degree as possible, consumers will receive value for their investment in the mental health programs provided.

The inclusion of fundraising efforts as a component of the strategic plan reflects not only the need to develop and update on a continuing basis a solid business plan in order to sustain Hopewell as a therapeutic care farm but also financial concerns resulting from the downturn in the allocation of funding for mental health care in general by the State of Ohio. In FY2009, Ohio's per capita spending on mental health care placed the state within the lowest 13 of the 50 states (National Alliance on Mental Illness, 2015, Appendix IV). Spending levels deteriorated further; between FY2009 and FY2012, mental health funding in the state of Ohio was reduced by $5.6 million, representing a 5.1 % decrease in available funding (National Alliance on Mental Illness, 2015, Appendix III).

References

American Psychiatric Association. (2013). *Diagnostic and statistical manual of mental disorders, fifth edition (DSM-V)*. Washington, DC: Author.

CARF International. (2015a). *Quick facts about CARF*. http://www.carf.org/About/QuickFacts/. Accessed 9 Apr 2015.

CARF International. (2015b). *Why does accreditation matter?* http://www.carf.org/Accreditation/. Accessed 9 Apr 2015.

National Alliance on Mental Illness. (2015). *State budget cuts report: State mental health cuts, the continuing crisis*. http://www2.nami.org/Template.cfm?Section=state_budget_cuts_report. Accessed 21 June 2015.

Ruch, D., & Carlton, C. (n.d.). Hopewell—A therapeutic farm community: Nature program. Presentation. Available from C. Carlton, ccarlton@hopewell.cc.

Chapter 5
CooperRiis Healing Community

The Establishment of CooperRiis Healing Community

History and Mission

Through the efforts of the founders Don and Lisbeth Riis Cooper, CooperRiis Healing Community (CooperRiis) was established as a private, not-for-profit organization and opened its doors on June 15, 2003. The Riis Coopers had first learned about therapeutic healing communities through their daughter's experience at Spring Lake Ranch, located in southern Vermont. Impressed with the underlying philosophy and programming at Spring Lake and a sister organization Gould Farm in Massachusetts, they sought to create additional similar opportunities for adults facing mental health struggles.

CooperRiis consists of a farm, located in Mill Spring, North Carolina; a dorm-like building, known as 85Z, fashioned from what was once a private mental hospital, located in Asheville, North Carolina; and 11 houses located in various parts of the greater Asheville community. The farm consists of 95 acres and includes two administration buildings; three lodges, each of which houses 12 residents; several houses for staff who live on-site; vegetable gardens; farm animals, including horses, goats, rabbits, and chickens; cats and dogs; a labyrinth; a pond; tennis courts; and numerous nature trails. The building known as 85Z provides a more urban setting with essentially the same services, with the exception of activities involving the animals. In total, CooperRiis has a staff of approximately 190 persons (170 full-time equivalents) with the capacity to provide services to 120 residents.

© Springer International Publishing Switzerland 2016   69
S. Loue, *Therapeutic Farms*, SpringerBriefs in Social Work,
DOI 10.1007/978-3-319-13539-7_5

CooperRiis describes itself as "a place of integration and recovery of wholeness, a place where isolation is replaced by relationship, and where the mind, body and spirit are nurtured within nature" (CooperRiis Healing Community, 2014b, p. 1). Its approach is clearly reflected in its mission statement:

> Our mission is to help individuals whose lives are impeded by mental health challenges or emotional distress to develop their capacities for creativity, wholeness, relationship, and optimal health so that they may achieve their highest levels of fulfillment and functioning and respond productively to their future challenges and opportunities for growth. (CooperRiis Healing Community, 2014b, p. 1)

The logo—a sunflower—reflects the community's reliance on nature as a therapeutic modality (Fig. 5.1).

Staff and residents function together as one integrated community, engaging in work, meals, and socialization together. Staff members are expected to reflect and foster:

Fig. 5.1   CooperRiis logo

- A culture of emotional and physical safety
- A personal life of meaning, belonging, hope, and purpose
- An atmosphere of kindness, respect, and open and authentic communication
- Each individual's ideas, contributions, and service to the work and the social life of the community
- Interactions that facilitate comfort and ease within the CooperRiis environment
- Relationships that are founded on empathy and mutuality (CooperRiis Healing Community, 2014b, p. 2)

Understanding Recovery

Phases of Recovery and Levels of Care

Recovery is conceptualized of, first, as a process rather than an event and, second, as a restoration of one's sense of identity and ability to enjoy a full and meaningful life, however the individual may define it for him- or herself. As such, recovery entails significantly more than the elimination or control of symptoms. This understanding of recovery can be analogized to diabetes, which often requires that the individual learn how to live a healthy life and not merely how to control his or her sugar and insulin levels. Accordingly, just as with diabetes, there is no cure for the condition, but the individual may recover and move forward from the symptoms that he or she is experiencing and the ill effects of those symptoms (Fig. 5.2).

Fig. 5.2 Recovery Rocks

Recovery is seen as progressing through three basic phases: initial, middle, and the third phase. However, because recovery is not a linear process, individuals may not move neatly from one phase to the next in a lock-step progression. Rather, the individual may move into a later phase and then again into an earlier phase, gradually making progress toward a fuller recovery.

The initial phase is characterized by the individual's acceptance of his or her mental health challenges and a desire to move forward. This phase often follows a long period of time during which the individual has been in denial of his or her mental challenges (CooperRiis Healing Community, 2014b, p. 7).

The second phase of recovery is focused on the individual's development of a sense of empowerment, his or her return to healthy functioning, and his or her adaptation to society (CooperRiis Healing Community, 2014b, p. 7). The individual's sense of identity may have disintegrated during the more acute phases of his or her mental illness; during the middle phase of recovery, the individual may redefine who he or she is as a person. The development of healthy behaviors, such as a regular sleep cycle, exercise, and healthy eating, is critical if the person is to progress to the final phase of recovery.

The third phase of recovery focuses further on the development of a stable sense of self and of well-being and positive attitudes. The individual continues to learn how to manage the symptoms of his or her mental illness, while developing a more defined sense of meaning and purpose (CooperRiis Healing Community, 2014b, p. 8).

Programming at CooperRiis is designed to facilitate individuals' development through each of the three phases of recovery. During each phase of recovery, programming is provided that focuses on life issues falling within the seven distinct domains comprising the enhanced recovery model. These seven domains include (Fig. 5.3): community and connectedness, spirituality, physical wellness, emotional and psychological health, purpose and productivity, empowerment and independence, and the domain of intellectual functions, learning, and creativity. Table 5.1 provides a description of each of these domains of recovery (CooperRiis Healing Community, n.d.).

Fig. 5.3 Recovery goals on wall plaque

The level of support provided to residents roughly corresponds to their phases of recovery. All individuals enter at Level A of care. At this level, during which individuals are supported 24 h per day by staff and the therapeutic milieu, residents are expected to live, work, and eat on the CooperRiis property (CooperRiis Healing Community, 2014b, pp. 3–4).

When a resident is ready to transition to community living, he or she may progress to Level B care (Transitional Housing). At the Level B of care, the resident may live in one of CooperRiis' community homes. In this setting, the resident is

Table 5.1 Description of seven domains of recovery

Domain	Description
Social/community/connectedness	Ability to connect to others in a healthy, balanced, and functional manner; to maintain appropriate boundaries; to develop authentic emotional intimacy and effective interpersonal skills; development of sense of pride, acceptance, and belonging
Spirituality	Development of a sense of inner peace and harmony, hopefulness, a feeling of appreciation and gratitude, clarity of values and purpose in life
Physical wellness	Ability to provide for one's own basic physical needs, e.g., hygiene, sleep, diet, exercise, and ongoing efforts to adopt a healthy lifestyle
Emotional and psychological health	Development of emotional stability and a sense of well-being; ability to self-monitor, self-soothe, and self-nurture; willingness to seek help; medication optimization
Purpose/productivity/fulfillment	Development of a sense of accomplishment and fulfillment derived from participation in meaningful and rewarding activities
Empowerment and independence	Development of attitudes and behaviors that foster increased independence and control over one's life and one's recovery process. Includes the ability to self-monitor and to utilize healthy coping strategies
Intellectual/learning/creativity	Willingness and desire to take healthy risks and challenge oneself mentally and creatively

supported by 24-h live-in staff and is provided with staff support for life and work goals, social activities, meal planning, budgeting, and transportation, while the resident is able to move freely within the larger community. The resident is also provided with weekly psychotherapy, access to a psychiatrist, regular meetings with a Recovery Coordinator to assist with case management needs and, as needed, supervision with self-administered medications and college and/or job coaching. Readiness to transition to the Level B of care is determined based on an assessment of various factors, including the individual's understanding of and ability to self-administer medications and his or her full engagement in any necessary clinical treatment, consistent upkeep of personal space and hygiene, ability to manage a budget, commitment to physical wellness, consistent commitment to community work and service, completion of designated groups, development of a Revised Distress Plan, and development of a Post-Transition Plan in collaboration with the Resident Recovery Team.

Level C care (Supported Housing) is similar to Level B care, but at this level, there is an increased focus on the residents' ability to live and work independently and there is no live-in staff. Level D care (Extended Community Services or ECS) is available to individuals who are able to live in their own housing with a limited amount of staff support, which may include access to a psychiatrist and to recovery-oriented social activities. Regardless of the level of care received, all residents have access to support services through an on-call system on a 24-h basis.

The average duration of stay at each of Levels A, B, and C is 6–9 months; the maximum length of stay at any of Levels A, B, or C is 3 years. There is no maximum length of stay at Level D. A limited number of individuals who require long-term individualized support and are unable to reach the levels of independence required for Levels B, C, and D may receive longer-term support through a "special arrangement" which attends to housing and personalized staff support.

Programming

All programming is designed to promote residents' recovery within each of the seven domains that are most relevant to them as they progress through the various phases of recovery. The Resident's Recovery Plan is focused via their development of a Dream Statement, which helps them personally to align their goals with the holistic richness of the therapeutic community. Table 5.2 provides an overview of the program components at each care level and the domains to which each program component relates.

Each day at CooperRiis begins after breakfast with a meeting of the staff and the residents in the meeting lounge of the main building. Each morning's meeting begins with staff and residents sharing their thoughts and feelings about a designated theme. As an example, Monday morning meetings begin with each resident and each staff member expressing their gratitude for something in their lives.

Table 5.2 Program components at A, B, C, and D levels of care

Level of care	Program components	Program elements/description	Associated domain
A	Individual psychotherapy	45–50 min/week May include cognitive behavioral therapy, interpersonal therapy, dialectical behavior therapy, trauma-based therapies, process experiential therapy, psychodynamic therapy, narrative therapy, neurofeedback and neuro-enhancement therapy, and/or art therapy, depending upon individual needs	Emotional and psychological health
	Neuro-enhancement therapy	Cranial electrotherapy stimulation (CES), which uses weak electrical currents, and transcranial direct current stimulation (tDCS), which uses visual stimulation, both of which are believed to enhance cognitive processing, memory, mood regulation, and impulse control. Not included in basic program; additional fee required	Emotional and psychological health
	Recovery planning and coordination	Regular meetings with Recovery Coordinator to assist in implementation of recovery plans; communication facilitation with family depending upon resident's needs and preference; coordination of services with outside professionals	Social/community/connectedness Purpose/productivity/fulfillment Empowerment and independence
	Group therapy	Residents are assigned to a group upon admission. Possibilities include dialectical behavior group, dual-diagnosis anonymous group, hearing voices group, newcomers group, process group, continuing recovery group, spirituality group, sexuality group, social awareness group, WRAP (Wellness Recovery Action Plan) group, wellness education group, transition group, men's and women's group	Emotional and psychological health Social/community/connectedness
	Substance abuse support	Intensive support	Emotional and psychological health Physical wellness
	Family support and education	3-day family education program; family therapy	Social/community/connectedness Empowerment and independence
	Medication optimization	Includes recovery-based psychosocial supports and services, sensitive and collaborative initiation of medication protocols, judicious medication tapering or withdrawal protocols, nutritional support, regular reassessment of recovery status to guide shared decision-making to adjust medication treatment; medication review at admission and every 4 weeks thereafter	Emotional and psychological health Physical wellness
	Physical health support strategies	Review of diagnosed physical conditions; referral to outside practitioners as needed	Physical wellness

	Service	Description	Domains
	Wellness planning and education	Completion of wellness assessment; assessment for smoking cessation support, as appropriate; dietary supplementation as prescribed by psychiatrist; recommendations for complementary health strategies such as yoga, massage and bodywork therapies, acupuncture, and/or meditation	Physical wellness Emotional and psychological health
	Residential arts and recreation	Instruction and materials for woodworking, pottery, painting, sculpting, music, and theatrical performance; recreational outings	Intellectual/learning/creativity Purpose/productivity/fulfillment
	Community work and service	Voluntary non-compensated work and community service 20 h/week At the farm, crews are devoted to six functions: campus, gardening, animal care, kitchen, woodshop, and ground maintenance. At 85Z, residents participate in On-Campus Crew, responsible for kitchen, campus, maintenance, gardening, and special projects, or the Off-Campus Crew, which volunteers at designated sites in the Asheville area	Social/community/connectedness Empowerment and independence Purpose/ productivity/fulfillment Intellectual/learning/creativity
B, C, D	Medication optimization assistance	Staff observation of self-medication at Level B and Number Nine; assistance obtaining medications from pharmacy; provision of information about medications	Emotional and psychological health Physical wellness Empowerment and independence
	Meal planning assistance	Assistance with meal planning; nutritional and cooking guidance at Level B and Number Nine and as needed at Levels C and D	Physical wellness Empowerment and independence
	Personal support	Assistance in maintaining proper personal hygiene and cleanliness/orderliness of living space	Physical wellness Emotional and psychological health
	Physical activity support	Assistance and support with exercise-related decision-making	Purpose/productivity/fulfillment
	Employment/education assistance	Assistance with seeking and retaining employment and/or educational opportunities	Emotional and psychological health
	Revision of Overall Recovery Plan	Assistance with the revision and refinement of goals and recovery plans	Purpose/productivity/fulfillment Empowerment and independence Intellectual/learning/creativity
	Rule enforcement	Oversight to ensure that residents are not engaged in self-harming activities or activities that may harm others; ensures abstinence from alcohol use at Levels B and C and Number Nine and illegal drugs at all levels	Physical wellness Emotional and psychological health
	Scheduling/service coordination	Assistance with scheduling appointment for psychiatry and/or psychotherapy, as needed; coordination of access to Level A respite care, as needed; coordination of routine medical and dental appointments; assistance planning social activities; transportation coordination, including NA/AA as needed	Emotional and psychological health Physical wellness Social/community/connectedness

Fig. 5.4 Activity board in gathering/meeting lounge, main building

Sharing is followed by any announcements and, after the close of the meeting, by
participation in various groups and work groups throughout the day, with a break for
lunch. Residents may also participate in activities following dinner or enjoy free
time (Fig. 5.4).

Outcomes

CooperRiis' dedication to the integration of the recovery model with evidence-
based practice at all levels of care and the continuous assessment of programmatic
outcomes is particularly noteworthy. Evidence-based practice, frequently referred
to as the medical model, often emphasizes illness, weakness, and the individual's
limitations in lieu of his or her potential for growth beyond the reduction or elimination of illness symptoms (Frese et al., 2001; Munetz and Frese, 2001). It has
been suggested that this medical model is responsible for the creation of a cycle
of disempowerment and despair due to its focus on the biology of mental illness
and the use of medications to reduce illness symptoms, sometimes using medication against the wishes of the individual (Frese et al., 2001; Munetz and Frese,
2001). Frese and colleagues have suggested how the medical model might be

reconciled with the recovery model to yield positive outcomes for the individual with mental illness:

> For persons who are so seriously impaired in their decision-making capacity that they are incapable of determining what is in their best interest, a paternalistic, externally reasoned approach seems not only appropriate but also necessary in most cases for the well-being of the impaired individual. However, as these impaired persons begin to benefit from externally initiated interventions, the locus of control should increasingly shift from the treatment provider to the person who is recovering. As individuals recover, they must gradually be afforded a larger role in the selection of treatments and services. Throughout the recovery process, persons should be given maximal opportunity to regain control over their lives. They should be given increasingly greater choice about evidence-based interventions and other available services. (Frese et al., 2001, p. 1464)

The commitment to understanding the effectiveness of the services provided to residents is evident at the levels of both the CooperRiis staff, including the director, and the board of directors/trustees. The director is required to provide board members "timely reports on comprehensive resident results studies and measurement of resident outcomes, applying methodology standards that give adequate results upon which to base reasoned decisions" (CooperRiis Healing Community, 2014a, p. 13).

Research conducted to date supports the view that recovery from the effects of mental illness is a complex and multifaceted process that requires a holistic, flexible, and individualized approach (Snyder, Schactman, & Young 2015). A recent assessment of resident outcomes using standardized measures indicated that, in general, residents continued to experience improvement across all domains of recovery and at all levels of care (CooperRiis Healing Community, 2013).

Additionally, on an annual basis, the entire CooperRiis community of staff and residents participates in a dialogue to rate the effectiveness with which CooperRiis addresses specified items, with the expectation that responses may serve as the basis for action items. Dialogue focuses on six questions:

1. Does CooperRiis provide a culture of safety, physical and emotional? Do you feel safe? Can you trust others?
2. Are you helping to meet CooperRiis' mission in a manner that enhances or sustains your sense of meaning; does your daily contribution help you to feel a sense of belonging, hope, and purpose?
3. Does CooperRiis foster an atmosphere of kindness, respect, openness, and authentic communication?
4. Is your participation in the social, recreational, and therapeutic life of CooperRiis welcomed and appreciated? Are you believed in? Does your "vote" make a difference?
5. Are you encouraged to feel at ease in the physical environment of CooperRiis? Does the quality of the dining experience also help you to feel at home and at peace?
6. Mutuality: are you empowered and encouraged to assert yourself and plan for your future, while also being encouraged to feel empathetic for others and supportive of them to so the same (CooperRiis Healing Community, 2014b, pp. 37–38)?

References

CooperRiis Healing Community. (2013). *Outcomes research summary*. Mill Spring, NC: Author.
CooperRiis Healing Community. (2014a). *Board operating principles*. Mill Spring, NC: Author.
CooperRiis Healing Community. (2014b). *Manual of policies, programs, procedures and practices*. Mill Spring, NC: Author.
CooperRiis Healing Community. (n.d.). *Description of CooperRiis recovery domains*. Mill Spring, NC: Author.
Frese, F. J., III, Stanley, J., Kress, K., & Vogel-Scibilia, S. (2001). Integrating evidence-based practices and the recovery model. *Psychiatric Services, 52*(11), 1462–1468.
Munetz, M. R., & Frese, F. J., III. (2001). Getting ready for recovery: Reconciling mandatory treatment with the recovery vision. *Psychiatric Rehabilitation Journal, 25*(1), 35–42.
Snyder, M., Schactman, L., & Young, S. (2015). Rates and correlations of change in three dimensions of recovery within a recovery model oriented therapeutic community. *Psychiatric Quarterly, 86*, 123–136.

Chapter 6
Slí Eile

The Establishment of Slí Eile

Slí Eile sits in Burton Park in the village of Churchtown, Ireland, approximately a 1 h drive from the larger city of Cork. The farm itself has been in existence for approximately 500 years, but Slí Eile was established on its premises only a short 2 and 1/2 years ago. The establishment of the farm followed significant political struggles with both politicians and local community members in 2006 that followed Slí Eile's efforts to implement a pilot project in Charleville, just 10 miles or so from the farm. The farm and farmhouse sit on 50 acres that are rented from the farm's owner; an additional 90 acres are reserved for forestry. Slí Eile also has a bungalow in the nearby town of Charleville, a portion of which has been transformed into a bakery that, together with the farm, serves as sites for Slí Eile residents to develop and improve the socialization and occupational skills needed to interact with coresidents and the world outside of Slí Eile.

Slí Eile was established as a voluntary nongovernmental organization and is registered as a charity. The organization actually consists of two separately regulated limited companies, one dedicated to the provision of social housing, affiliated with the Irish Council for Social Housing, and the second dedicated to the provision of social support services, funded through mental health services. This organizational structure somewhat follows the model of a care focus farm, which provides care through collaborative relationships and governmental support. (See Chap. 3 for further detail about this model.) The enunciated goal of Slí Eile "is to provide a safe environment conducive to recovery" (Sapouna, 2007, p. 6). As such, Slí Eile's approach is congruent with the goals of the Mental Health Commission to ensure that "peer support/advocacy is available to service users" and that "service users experience a recovery-focused approach to treatment and care" (Mental Health Commission, 2007, Standards 3.3, 3.5, pp. 32–22).

That Slí Eile exists is remarkable in view of the many obstacles that faced its founders. Slí Eile, meaning "another way," was the product of Joan Hamilton's

© Springer International Publishing Switzerland 2016   79
S. Loue, *Therapeutic Farms*, SpringerBriefs in Social Work,
DOI 10.1007/978-3-319-13539-7_6

search for an answer to her daughter's mental illness and the apparent failure of Ireland's mental health-care system to help her. Slí Eile represents Hamilton's third attempt to establish such an organization. (For a short video clip describing efforts to establish Slí Eile, see the trailer for "Another Way Home," a film shown at the 2012 Corona Cork Film Festival, http://www.completecontrolfilms.com/another-way-home/.) Efforts by Hamilton and her board of trustees to establish a facility at Pike Farm in Charleville met with significant opposition from the local populace following the disclosure of the organization's focus on mental illness by a local politician (Anon, 2005; Browne, 2005). Slí Eile founders were confronted on a continuing basis with demonstrators bearing placards that demanded "Slí Eile Get Out — Stay Out" and "No Support No Project Get the Message" (Interview with Harry Gijbels, Chairperson, Board of Directors, 10 July 2015, Cork, Ireland; Anon, 2005; Another Way Home/Trailer, 2012; Browne, 2005).

Difficulties with the neighbors remained irresolvable at that time, and in 2006, Slí Eile was offered an alternative property, the Villa Maria, in Charleville. Villa Maria initially housed five tenants. Individuals were accepted into Villa Maria only if they had some ability to live independently and engage socially, since only a low level of support was to be provided. Days were structured around the tasks of daily living, such as cooking and cleaning, and work at the bakery. (For a short video featuring the residents, their experience at Villa Maria, and their work at the bakery, see Slí Eile's "Behind the Walls," https://www.youtube.com/watch?v=8DNJOKbk07w.)

Villa Maria's first tenants appeared, in general, to benefit from their experience. The tenants themselves reported improvements in their ability to communicate and build friendships, to voice their own opinions, and to participate in decision-making (Sapouna, 2007, pp. 7–9). Additionally, a study by the mental health services of Slí Eile's delivery of services was apparently resolved favorably, and the funding was continued (A Vision for Change Monitoring Group, 2012; Office of the Assistant National Director & Mental Health, 2011). The assumption of personal responsibility for the tasks of daily living and the need to adhere to a structure proved to be too difficult for some tenants whose lives had been defined by chaos and for those who had learned to remain dependent on others for their maintenance. For still others, life at Villa Maria did not offer enough.

Quite serendipitously, Hamilton learned that the farm in nearby Churchtown was to become vacant and was able to negotiate a 10-year rental agreement with the owner beginning in 2012. The farm currently houses seven residents and has the potential, depending on financing, to house an additional seven. The bungalow that adjoins the bakery in Charleville is sufficiently large to provide unstaffed housing for several additional individuals wishing to move on from the farm. Residents of Slí Eile, referred to as tenants, finance their rent and board from the disability allowance that they receive from the government. Individuals sign both a tenancy agreement that details their rights and obligations as tenants and the obligations of Slí Eile and an agreement that permits direct payment by the bank to Slí Eile's rent account. (See Appendix for a copy of the Tenancy Agreement, available at http://www.slieile.ie/wp-content/uploads/2011/09/tenancy-agreement.doc.) Daily expenses of food, heating, and so forth are met through tenant contributions to a

Fig. 6.1 The Slí Eile farmhouse

household fund that is managed by the tenants themselves, with the support of staff. Slí Eile is staffed by nonclinical personnel: a full-time director, one full-time support worker, and five part-time support workers, in addition to a full-time service manager. There are also part-time managers for horticulture and for livestock and care of the 50 organic farming acres.

Programming

Individuals seeking residence at Slí Eile meet with Hamilton to discuss their individual situation, including their needs and their finances. Following that meeting, which always occurs on a Wednesday, prospective tenants are asked to notify Hamilton within 2 days if they remain interested in joining the Slí Eile community. If an opening exists, they are invited to Slí Eile for a 2-week trial period, during which the individual and the existing Slí Eile tenants and staff mutually assess the suitability of the prospective tenant and Slí Eile for each other.

Unlike the therapeutic farm communities described in previous chapters, Slí Eile does not provide direct care services. Tenants continue to obtain mental health services from their mental health team that was in place prior to their residence at Slí Eile and attend clinical appointments with support from Slí Eile staff. The community utilizes William

Table 6.1 Table of the Seven Caring Behaviors and
the Seven Deadly Behaviors of Glasser's choice theory

Caring behaviors	Deadly behaviors
1. Listening	1. Blaming
2. Encouraging	2. Complaining
3. Negotiating	3. Criticizing
4. Supporting	4. Bribing
5. Trusting	5. Threatening
6. Affirming	6. Punishing
7. Respecting	7. Nagging

Glasser's choice theory and reality therapy as the basis for its approach. Choice theory postulates that humans choose most of their behaviors and that genes drive humans' efforts to fulfill their five basic needs of survival, love and belonging, power, freedom, and fun (William Glasser Institute, 2010a). Staff and tenants are encouraged to practice the Seven Caring Behaviors and refrain from engaging in the Seven Deadly Behaviors of Glasser's choice theory (William Glasser Institute, 2010a) (see Table 6.1).

Behavior is analogized to the operation of a car; the front wheels represent thinking and doing and the back wheels of the car signify feeling and physiology. Together, the front and back wheels comprise total behavior. Tenants are encouraged to rely increasingly on the front wheels and less on the back wheels as they make decisions about their behavior and seek to meet the five categories of basic needs identified by Glasser. (This focus on thinking bears some resemblance to cognitive behavior therapy, discussed in Chap. 2.)

According to Glasser, reality therapy derives from choice theory. Reality therapy urges individuals to focus on the present rather than the past; avoid discussing symptoms and complaints; focus on what can be accomplished through thinking and acting; refrain from criticizing, blaming, or complaining; be nonjudgmental and noncoercive; avoid allowing excuses to interfere with their ability to make connections; attend to specifics; and develop specific, achievable plans with appropriate follow-through (William Glasser Institute, 2010b).

Individuals joining the Slí Eile community progress through four stages. During the first stage, individuals are asked to self-evaluate their ability to understand and adhere to any medication regimen that may have been prescribed by their independent psychiatrist. During stage 1, individuals are asked to evaluate on a daily basis their ability to accomplish basic tasks of daily living, such as making their beds and arriving for breakfast on time. During this stage, tenants are provided with six individual counseling sessions with a therapist who works with Slí Eile as an independent contractor (see Fig. 6.2). Tenants are also tasked with the responsibility of evaluating their own behavior, the consequences of their behavior, and alternative approaches to dealing with various situations that may arise. (See Table 6.2.)

Beyond stage 1, the skills needed to move to the next stages are not as well defined. Individuals are able to progress to stage 2 when they are able to independently monitor their medication adherence. At this stage, individuals maintain

---

Guide for Tenants on their Journey with Sli Eile

STEP 1

Respect

- Does my behaviour reflect self-respect and respect for others

Responsibility

- Do I understand the purpose of Cuisine Slí Eile?
- Am I familiar with the William Glasser approach used in Sli Eile
    - ✓ Understanding the CAR,
    - ✓ Familiar with 7 good habits & 7 deadlies
    - ✓ Have watched the Glasser DVD
- Am I prepared to take responsibility for having my medication in the locked drawer in my bedroom?
- When I get up in the morning, am I
    - ✓ making my bed
    - ✓ opening my window
    - ✓ taking care with personal hygiene
    - ✓ wearing clean clothes
    - ✓ signing my medication record (if taking medication)
    - ✓ arriving for breakfast on time
    - ✓ when on cooking, do I get up earlier, having prepared porridge & fruit previous evening
- Am I recording and achieving short term personal goals
- Am I completing my personal washing on my designated day
    - ✓ Washed
    - ✓ Dried
    - ✓ Aired
    - ✓ Folded in my bedroom
- Have I completed my 6 counselling sessions

Fig. 6.2 Guide for tenants' self-evaluation, step 1

responsibility for a week's worth of their medications and place their order for a new supply 48 h before their supply is finished. They assume increased responsibility for household chores, as well as responsibility for a personal budget. Individuals who progress to stage 2 are expected to achieve their short-term goals that they identified during step 1 and continue to develop additional goals.

At stage 3, individuals begin to develop more specific goals for employment, education, and/or their living situations, which they then put into action during stage 4. Additionally, they act as a buddy for new tenants and participate in the planning of the Slí Eile tenants' meals (Hamilton & Gijbels, 2014). It is expected that individuals who have progressed to stage 4 will have developed goals that include participation in some activities off-site, such as school attendance, part-time employment, and/or volunteer services, and that these goals will be acted upon during stage 4.

Table 6.2 Questions for self-evaluation

1. Is it working?
2. Is it helping?
3. How is it going?
4. If you continue to act that way, how will you feel?
5. Is this the direction you thought you would take?
6. What do you have to change before you feel more comfortable?
7. Are you doing better? Feeling better? Getting better?
8. Which one do you think would be best to work on first?
9. Are you satisfied with the way things are?
10. What would you have to change, so things would improve?
11. Is it better or worse now?
12. Can you get there from here?
13. In what ways have things improved?
14. What indicators would you look for that would tell you that you're making progress?
15. Would practicing this help?
16. Would it be more helpful to you if this was in writing?
17. Do you need some help?
18. Are you getting what you want? If not, what could you do differently?
19. Are your needs being met?
20. If this path is not taking you in the direction you want to go, what path might be more
    effective?
21. What do you have to do so things will get better?
22. Are you willing to work this out?
23. Is what you're doing working?
24. Who's in control when you feel upset?
25. Are you choosing this?

Work itself is said to serve as the therapy ("therapeutic engagement"). Days are highly structured, beginning with breakfast at 7 a.m. Tenants leave for the bakery at 7:25 a.m., where they prepare the goods that have been ordered by local businesses. Once the baked goods—scones, breads, pies—have been prepared by the tenants together with a full-time support worker, the tenants head off to deliver them to the various merchants (Fig. 6.3).

Proceeds from the bakery are used to sustain the bakery enterprise and to cover six individual counseling sessions for each tenant during stage 1 and one-half of the cost of individual counseling sessions requested by a tenant beyond the six stage 1 sessions. (The tenant pays for the remaining one-half cost from their government-provided stipend.) All new tenants begin their Slí Eile experience as part of the bakery team but, on an individual rotating basis, remain at the farm rather than heading to the bakery. On reaching stage 3, they leave the bakery team and remain at the farm assisting staff in activities related to horticulture, general farm maintenance, animal care, or housework (Figs. 6.4 and 6.5).

The day's main meal is at lunchtime, followed by a community meeting, during which the tenants and staff discuss the events of their previous day. Community meetings with all tenants and staff are held on a weekly basis to provide tenants and staff alike with an opportunity to discuss any changes within Slí Eile, to participate

Fig. 6.3 (a, b) Bakery orders and baked goods

Fig. 6.4 The fields

Fig. 6.5 New piglets

in decision-making at the community level, to update and arrange for prescriptions, and to discuss the "comings and goings" within Slí Eile, such as informing everyone about clinical appointments and upcoming visitors to the farm and providing details about farm events (Hamilton, n.d.).

Tenants also have free time to engage in self-directed and selected activities. These may include hiking, reading, writing, visiting with family members and friends, and shopping. Volunteers provide tenants with additional opportunities for interaction and socialization. Some residents utilize their free time to write creatively. One resident has authored numerous poems, two of which have been published in the local newspaper and another of which appears below.

Sunflower Harvest in the Farm

I stay with the volunteer gang wary of the undergrowth
Darkened by 5-foot high sunflowers in the sunflower meadow.
We cut the flowers, 2000 between us all
To sell in the wheelbarrow soon, in the city. It is evening
On leaving your wife sees me, pulls me toward her
In cuddles, the same look on her face as yours; it is time to go.
Check grey shirt and grey trousers, camouflaged against
Silver brick sheds the silver brick arch
That outlines a path to the sunset. Check silver shirt, silver trousers
Decorate the grey-brick sheds and grey brick arch with the reason you came.

~ Anita

Outcomes

Slí Eile has not systematically examined the outcomes of its approach. To some degree, such an assessment would be difficult as there are no standard metrics for tenant assessment when they enter the program and no standard metrics for tenant assessment when and if they leave. No time limit is imposed on the duration of an individual's residence at the Slí Eile farm. In the longer term, the lack of metrics and an assessment may have financial implications. As resources become increasingly scarce, funders may be wary of providing support to a venture whose outcome is not sufficiently substantiated by data, whether those data are qualitative or quantitative in nature.

Nevertheless, a progress report compiled by a researcher with the Department of Applied Social Studies of University College Cork indicated on the basis of interviews with residents of then-functioning Villa Maria that the residents obtained benefits from the program. Residents reported having a newfound sense of belonging to a group, building friendships, increasing their level of self-confidence, developing a sense of collaboration and ownership, being able to make plans for the future, and experiencing a sense of safety (Sapouna, 2007, pp. 7–16). Many of the report's recommendations for program improvement have been implemented since the time of that report, suggesting that quality improvement is a high priority for the organization. These program modifications include increasing tenant participation in the evaluation of their own progress, creating opportunities for joint planning of activities between tenants and staff, involving tenants and staff in broader decisionmaking, providing increased opportunities for staff and tenants to discuss tenants' future goals, providing tenants with greater clarity regarding their rights as tenants (see Appendix for current tenancy agreement), instituting regular tenant-staff house meetings, and increasing staff and tenant participation in the referral process.

Looking Toward the Future

Slí Eile's tenure on the farm remains somewhat uncertain due to both financing and potential sale of the farm by its owner. The farm is in need of significant repair but, because Slí Eile is a tenant rather than an owner, the investment of substantial monies for repairs at this time is inadvisable.

Despite this uncertainty, the director and the board have engaged in strategic action planning to plan for the future. Currently contemplated endeavors for the upcoming 5 years include maintaining the option of a day program for users of other services, e.g., individuals on the waiting list for a vacancy as a tenant at Slí Eile; the establishment in July 2015 of a farm gate outlet for the sale of the farm's organic produce to include a small café for the sale of tea, coffee, and the bakery's pastries; the opening of a small dairy in 2016 to process the farm's organic milk and to develop additional products such as yogurt, butter, and soft cheese; the hiring of a full-time administrator; and transition of Slí Eile to a new director following Hamilton's retirement.

Efforts are also being made to enhance the programmatic guidelines in order to provide additional guidance to tenants and staff alike. As an example, Hamilton is currently considering the development of specific competencies or skills to be achieved at each stage in order to provide both tenants and staff with better defined metrics for successful attainment prior to advancing to the subsequent stage. (See Appendix.)

Appendix

TENANCY AGREEMENT

Made the_____ day of _____ 200___

Between

Slí Eile Housing Association Ltd.

(Hereinafter called the 'Owner')

_____

(Hereinafter called the 'Tenant')

Whereas

The above-named Tenant has been accepted by the Owner as being in need of housing under the terms of the Housing Act 1988.

And whereas

The above-named Tenant hereby undertakes to occupy the accommodation allocated to him/her during the period of the tenancy and to peaceably surrender same to the Owner in the event of such accommodation ceasing or in case of any other breach, non-performance or non-compliance with the provisions, terms and conditions set out in this Agreement.

And whereas

The Owner, being the beneficial owner of the said dwelling as described in the First Schedule attached hereto, hereinafter referred to as the "dwelling", **agrees** to let and the Tenant **agrees** to take the accommodation for the period and subject to the rent and manner of payment thereof specified in the Second Schedule attached hereto.

(a) To pay the Rent specified in the First Schedule without any deductions in the manner provided for and not later than the day or date prescribed.
(b) To seek permission of the Owner in the event of the Tenant wishing to share his/her accommodation for any period of time.
(c) To permit reasonable access to the Tenant's personal accommodation by staff who have prior permission from the Owner.
(d) Not to make any structural alterations to the Dwelling without prior consent of the Owner.
(e) To take responsibility for maintaining a satisfactory level of personal hygiene and cleanliness throughout the Dwelling.
(f) Not to intentionally or carelessly damage any fixtures, fittings or appliances that are the property of the other Tenants or Owner.

(g) Not to do anything that is likely to be or become a nuisance, danger or annoyance to the Owner, other occupiers of the Dwelling or to occupiers of adjoining or adjacent buildings, gardens or amenity land.

(h) To seek the permission of the Owner before attempting to keep any kind of pet in or about the Dwelling.

(i) To agree to abide by the rules of the Dwelling as formulated through discussions with the other Tenants, staff and Owner. Such rules to be positioned in a place of prominence in the Dwelling.

(j) To contribute to the service charges and bills in respect of all essential services such as electricity, gas, water, refuse collection etc. as agreed payable according to Schedule 2 attached herewith.

(k) To participate in all day-to-day household duties including cooking, cleaning, laundry and the upkeep and maintenance of the Dwelling and garden.

(l) At the expiration or sooner termination of the tenancy to peaceably surrender same to the Owner, together with any furniture, effects, fittings and appliances included in this accommodation.

(m) To take responsibility for the cost of repair or replacement as necessary for any items or structures maliciously or carelessly broken or damaged by the Tenant.

AND THE OWNER AGREES WITH THE TENANT that the Tenant paying the rent and services charges and performing, observing or complying with the terms and conditions of this Agreement, hereinbefore contained, may peaceably hold and enjoy the dwelling during the period of the tenancy.

AND THAT the tenancy hereby created shall be one which can be terminated by either Tenant or Owner by 2 weeks notice in writing by either party to this Agreement.

AND THAT in the event of the Tenant requiring to be accommodated elsewhere for a period of time while, at the same time, wishing to retain the accommodation in the Dwelling, such accommodation will be permitted for a period up to 4 weeks providing that Rent continues to be paid for such period of time.

AND THAT the Owner shall endeavor to keep in good order and repair the structure, furniture, appliances and exterior of the dwelling, including plumbing, electricity, repairs etc. provided that same has not been maliciously or carelessly damaged or broken by the Tenant.

AND THAT in the event of the Rent and Services Charge, or any part thereof being in arrears for 7 days after becoming due (whether formally demanded or not) or if there is any breach or non-performance or non-compliance by the Tenant with any of the terms and conditions and covenants contained in this Agreement, the Owner shall be entitled to terminate the tenancy hereby created by serving on the Tenant 2 weeks notice in writing.

AND THAT the Tenant shall have reasonable access to the use of such common areas and facilities that are provided for the use of the Tenants both inside the dwelling and outside.

The Tenant and Owner ALSO ACCEPT AND AGREE TO COMPLY WITH THE FOLLOWING ADDITIONAL CONDITIONS AS APPROPRIATE

1. The Tenant and Owner agree that each person, whether living or working in the Dwelling, will be treated as a person worthy of dignity and respect.
2. All disagreements or difficulties will be resolved by the Tenants themselves in collaboration with agents of the Owner through the processes of facilitation, mediation or arbitration as appropriate.
3. The Tenant and Owner accept that the Dwelling is a place of personal recovery and growth, which encourages its occupants to achieve the necessary confidence and skills to move forward and take up their own place in society.
4. All rules and day-to-day living practices will be established and evaluated by the Tenants themselves with the assistance, as necessary, of agents of the Owner.
5. The Tenant undertakes to take responsibility for his/her own personal effects and belongings.
6. The Tenant and Owner agree to play their part, as appropriate, in the day to day activities involved in the running of the Dwelling e.g. cooking, cleaning, gardening, administration, shopping etc.
7. The Tenant and Owner agree to maintain the principles that are central to the success of this supported social housing project.
8. Each Tenant agrees to endeavor to cooperate fully in the achievement of his or her own pathway to independent living. The Owner agrees to endeavor to do everything possible to enable such achievement through the supports put in place.
9. The Tenant agrees to take increasing responsibility for his/her own appointments, medication, family commitments, choice of activities and training opportunities. The Owner agrees to share this responsibility to whatever level is individually appropriate.

FIRST SCHEDULE

DESCRIPTION OF LETTING

ALL THAT THE DWELLING COMPRISING
The sole occupation of one bedroom and the sharing, with other Tenants, house, conservatory, garden and work shed at:
Villa Maria, Smiths Road, Charleville, Co. Cork

Including any furniture, effects and fittings specified in the inventory List shown in the Rent Book and the Tenant accepts these are part of this letting. EXCEPTING AND RESERVING unto the Owner the exterior walls of the said dwelling and the free and uninterrupted passage of gas, water and electricity or television circuits through the pipes and wires and cables which now are or may at any time hereafter be in, on or about or passing through the said dwelling with access allowed at all reasonable times for the Owner or the Owner's agents or work persons to enter the dwelling for the purpose of inspecting, repairing, replacing or altering any such pipes, wires of cables.

SECOND SCHEDULE

Period of Tenancy, Rent, Service Charges and Manner in which payable

PERIOD: _____ TO _____

RENT CHARGE   € _____

SERVICE CHARGE
(Including electricity, gas, heating fuel, L.A. services, payphone, TV licence,   € _____
essential maintenance, upkeep and repairs)

MANNER PAYABLE   *(Tick as appropriate)*

1.   By Weekly Standing Order to a Bank Account designated by the Owner   _____

2.   By Monthly Standing Order to a Bank Account designated by the Owner   _____

Date in month or Day in Week on which Rent is payable: _____

NOTE: The above details should accurately correspond with the details given in the Rent Book supplied to the Tenant.

The Tenant shall be responsible for the maintenance of hygiene, tidiness and cleanliness in accordance with the levels agreed by all those occupying the dwelling at any given time.

Signed by the Owner: _____   Co. Secretary - Slí Eile H.A. Ltd.

Signed by the Tenant: _____
(Block capitals)_____

 In the presence of: _____   Date: _____

Slí Eile
Table of steps, goals, activities, and competencies

Step	Primary goal(s)	Activity/activities to achieve goals	Competencies to be developed
1	Development of basic aspects of self-understanding	• Completion of daily evaluation sheet • Identification of weekly goals • Participation in community feedback • Participation in communication group	Tenant will be able to: • Engage in self-evaluation, including understanding of own emotional responses • Identify triggers and responses to triggers • Communicate needs to others • Recognize need for support • Understand and apply car model to one's own behavior
2	Development of insight into needs for healthful living	• Participation in planning and preparation of healthful meals for community living in collaboration with support staff • Maintenance of room and common areas	Tenant will be able to: • Explain his/her identified needs to maintain healthful living, e.g., dietary needs, exercise regimen, medication regimen • Adhere to medication regimen with staff oversight • Plan and prepare healthful menus • Maintain a neat and orderly room • Participate with others in maintaining a clean home • Launder and iron own clothes • Maintain good personal hygiene
	Development of socialization and communication skills	• Work in bakery • Deliver baked goods to businesses *Activities to be performed with staff support and oversight*	Tenant will be able to: • Work collaboratively as a member of a team • Adhere to a structured work schedule developed by staff • Explain procedures and policies to new tenants • Communicate feedback to team members in a positive and constructive manner

(continued)

Slí Eile (continued)

Step	Primary goal(s)	Activity/activities to achieve goals	Competencies to be developed
3	Development of independent living skills	• Work on farm in agriculture-related work and/or animal care • Self-management of medication (prescription placement, blood tests, medication usage) in accordance with prescribed regimen *Activities to be performed independently and/or in collaboration with staff without supervision or oversight*	Tenant will be able to: • Identify, explain, and implement strategies for self-sustainability (scheduling, planning, budgeting) • Identify independently and without staff prompting the tasks that are to be accomplished in connection with agricultural work and animal care at Slí Eile • Plan and adhere to a schedule to achieve the required tasks • Identify anxiety-producing situations when functioning independently • Identify and utilize coping strategies to reduce stress and anxiety
		Identification of mid-term goal(s)	Tenant will be able to: • Identify strategies to accomplish mid-term goal(s) • Develop a realistic timeline for achievement of mid-term goal(s) • Identify potential challenges to accomplishment of mid-term goals and strategies to address challenges, e.g., revision of goal and/or timeline • Achieve identified mid-term goal(s) and/or revise • Formulate on a preliminary basis longer-term goal(s)
4	Development of ability to work and function socially in unsheltered environment	Work outside of Slí Eile (volunteer work, paid employment, study)	Tenant will be able to: • Identify strategies to accomplish longer-term goal(s) • Develop a realistic timeline for achievement of longer-term goal(s) • Identify potential challenges to accomplishment of longer-term goals and strategies to address challenges, e.g., revision of goal and/or timeline • Achieve identified longer-term goal(s) • Develop enlarged support network outside of Slí Eile

Note: Progress through these steps does not always occur as a linear progression, moving from one step to the next. Very often, the recovery process involves moving up to a step in some ways, and remaining at a previous step in other ways, or even going back to a previous step for a brief period of time. This is part of the normal recovery process. It is not a failure!

References

A Vision for Change Monitoring Group. (2012, June). *A vision for change—A report of the Expert Group on Mental Health Policy. Sixth annual report on implementation 2011.* http://health.gov.ie/wp-content/uploads/2014/03/vision_for_change.pdf. Accessed 20 July 2015.

Anon. (2005). Complaint made on minister's role in residential home siege. Politico: Social and political issues, May 27. http://politico.ie/archive/complaint-made-ministers-role-residential--home-siege. Accessed 14 July 2015.

Another Way Home/Trailer. (2012). M. Twomey, director; produced by Complete Control Films. http://www.completecontrolfilms.com/another-way-home/. Accessed 14 July 2015.

Browne, V. (2005). Reporting on Charleville protest. Politico: Social and political issues, May 27. http://politico.ie/archive/reporting-charleville-protest. Accessed 14 July 2015.

Hamilton, J. (n.d.). Induction handbook. Churchtown, Ireland: Slí Eile.

Hamilton, J., & Gijbels, H. (2014). Slí Eile Farm: A community living farm project for people experiencing mental health difficulties. In *INTAR [International Network Toward Alternatives and Recovery] Conference*, Liverpool, UK, June 27.

Mental Health Commission. (2007). *Quality framework: Mental health services in Ireland.* Dublin: Author. http://www.mhcirl.ie/File/qframemhc.pdf. Accessed 3 Aug 2015.

Office of the Assistant National Director, Mental Health. (2011). *Implementation of a vision for change in 2011.* http://health.gov.ie/wp-content/uploads/2014/04/HSE_National_Regional_Progress_Report.pdf. Accessed 20 July 2015.

Sapouna, L. (2007, May). *Progress report on the Slí Eile housing project.* Cork, Ireland: Department of Applied Social Studies, University College Cork.

Slí Eile. (n.d.). Behind the walls. https://www.youtube.com/watch?v=8DNJOKbk07w. Accessed 14 July 2014.

William Glasser Institute. (2010a). *Choice theory.* http://www.wglasser.com/the-glasser-approach/choice-theory. Accessed 16 July 2015.

William Glasser Institute. (2010b). *Reality therapy.* http://www.wglasser.com/the-glasser-approach/reality-therapy. Accessed 16 July 2015.

Chapter 7
Moving Forward: Exploring Current Challenges and New Directions

Therapeutic/care farm communities potentially offer participants numerous benefits and opportunities, depending upon their specific organizational model. These include the development and/or improvement of socialization skills; a structured environment that facilitates meaningful work; the acquisition of new skills that may provide the foundation for new vocational opportunities; the establishment and/or reinforcement of a support network; and that enhancement of skills required for successful daily living, such as budgeting, banking, and food preparation. These individual achievements may heighten participants' self-confidence and self-esteem and provide encouragement as the participants' progress further along their journeys to recovery.

Nevertheless, today's therapeutic farm communities face numerous challenges to their continued existence, much as the therapeutic communities of the past also faced. These challenges arise within the context of the relationship between the individual clients and the farms themselves and within the financial, legal, and political domains.

The Client-Community Relationship

Arguments have been advanced by some farms that individuals' potential for recovery is limited as the result of their diagnosis with a mental illness, essentially equating mere diagnosis with the "medical model." It has also been argued that the likelihood of recovery is reduced as a result of labeling and categorizing an individual's behaviors. It is suggested here that a rigid refusal to clinically assess behaviors and symptoms, sometimes framed as "leaving your diagnosis at the doorstep," raises questions about the adequacy of the support services and/or clinical services provided to the consumers/residents, particularly in the context of residential services for those who are more seriously ill. It is unlikely that such a perspective would be acceptable in the context of service provision to individuals with severe

© Springer International Publishing Switzerland 2016   97
S. Loue, *Therapeutic Farms*, SpringerBriefs in Social Work,
DOI 10.1007/978-3-319-13539-7_7

diabetes, epilepsy, or cardiac conditions. For example, in the case of illnesses such as these, staff would minimally be trained to recognize the warning signs and symptoms of a seizure or cardiac event and take minimal measures to ameliorate the situation to the degree possible while waiting for the emergency services. At the therapeutic farms where personnel have had no clinical training, there may be an increased likelihood that personnel—both the support personnel and those engaged in the farming enterprise—will be unable to recognize the signs of decompensation, distinguish between suicidal gestures and an actual suicide attempt, or de-escalate a potentially violent situation.

Some farms might also assert that the therapeutic farm is merely a place of employment, and, in that context, knowledge of an individual's diagnosis and/or behavioral symptoms is irrelevant and, in some jurisdictions, illegal unless disclosed voluntarily by the individual himself or herself. If the individual's experience is characterized as employment, one must ask why he or she is not receiving fair wages in exchange for the labor performed, an issue that is discussed further below. If the individual's experience is, however, characterizable as therapeutic, then the question must be raised as to whether knowledge and understanding of the diagnosis and associated symptoms is relevant and how such information can be utilized to help the individual recover from his/her mental illness to the maximum degree possible.

Frese, Stanley, Kress, and Vogel-Scibilia (2001) have suggested that both the scientific evidence-based approach, which relies on external scientific reality, and the recovery model, which emphasizes the subjective experiences and autonomous rights of recovering individuals, must be considered in making treatment decisions (Frese et al., 2001, p. 1463). Clearly, "evidence-based treatment may differ from treatments that are based on the recovery model insofar as they reflect different judgments of the value of various treatment outcomes by service providers and consumers" (Frese et al., 2001, p. 1463). In speaking of treatment, we can conceive of treatment as significantly broader than merely the administration of medication, encompassing the nature of the therapy(ies), the setting in which it is provided, the provider of the therapy, and other aspects of care.

Munetz and Frese (2001) considered the spectrum of mental illness in fashioning a balance between the medical/scientific and recovery approaches. They recommend that a paternalistic, scientifically premised approach be utilized for individuals whose decision-making capacity is so impaired as to obviate their ability to determine their own best interest. However, as individuals gradually recover from the effects of their mental illness, increasing control and choice of treatment should be ceded by the provider(s) to the recovering individuals so that the individual can regain the maximum degree of control over his or her life as possible.

Implementation of this scaled approach would seem to preclude the acceptance of seriously ill individuals by therapeutic/care farms that lack clinical expertise and provide low levels of support. Whether the determination of symptom severity and the level of support needed requires a clinical assessment or whether a judgment can or should be made solely on the basis of observation by nonclinically trained staff and other program participants requires careful and considered examination.

A related issue concerns the degree to which staffing should be professionalized. This issue is both internal to therapeutic farms and a question of the degree of regulation required by the specific jurisdiction in which a particular farm operates. Because the therapeutic farms in the United States often receive funding through health insurance reimbursements, whether through private health insurance or government-funded programs, a minimal allowable level of professionalization is required both by regulation and to merit reimbursement. The extent to which mental health credentials are or should be required of some or all staff appears to be a controversial and contentious issue in Europe. Rather than assuming an all-or-nothing approach, whereby professional credentials/licensing is required, it may be helpful to consider the following questions with regard to each individual therapeutic farm:

- Mental illness and distress exist along a spectrum of acuity and severity. How severe are the residents' symptoms? For example, are residents accepted if they are acutely psychotic? Do all residents have the capacity to engage socially and function somewhat independently? The more seriously ill the clients are, the stronger the argument is for clinical training of at least some staff members.
- Are clinical services provided by staff or by independent contractors engaged by the facility or are they obtained independently by the resident? Does the facility administration have the ability to evaluate the quality of the services provided and the level of benefit to the clients?
- Regardless of credentials, what training do staff need and receive to be able to recognize and address mental health and/or physical emergencies, e.g., a client's decompensation or drug overdose, the need for cardiopulmonary resuscitation (CPR)?

The therapeutic farms in both the United States and Europe generally view the work done by residents as therapy. Some, in addition, maintain that the farm component of the enterprise—the animals, the horticultural activities, and other farming-associated activities—must be sustainable financially through the work of the residents. However, it is likely that most, if not all, of the therapeutic farms in the United States and at least some of the farms in Europe do not pay wages to the individuals for their farm-associated labor, stemming from the premise that work is therapy. The nonpayment of wages raises ethical, and potentially human rights, issues regarding the possible exploitation of mentally ill individuals. Legal issues may also arise related to both the violation of minimum wage laws and laws against unfair competition due to the economic disadvantage that may be experienced by farms that pay for labor performed for their benefit. Some therapeutic farm communities have proactively addressed such concerns by creating employment training programs that provide for internship positions, followed by on-the-job training opportunities that rely on the foundational skills learned during the internship, and that culminate in the possibility of paid employment.

Depending upon their theoretical orientation, client population, and understandings of recovery and mental illness, some residential farm programs permit clients to reside at the farm indefinitely only in exceptional circumstances, while others place no limit on the duration of a client's residence as long as the client is able to pay the required fees. And, while some farms encourage or even require their clients to participate in programs that help them learn new marketable skills for their post-farm lives, others place greater emphasis on the sustainability, stability, and continuity of the community of residents, focusing less attention and fewer resources on fostering client abilities to function independent of the farm community residents and staff. These diverse approaches suggest the need for developing a balance between efforts to sustain the community of residents and efforts to encourage the development of individual autonomy. Indeed, loss of residents from a community can itself lead to trauma among some remaining residents and destabilization of the therapeutic community, in much the same way that loss of a family member may lead to trauma and destabilization of the family. It is this writer's opinion that an emphasis on maintenance of the community rather than the fostering of individuals' growth may potentially transform what should be an opportunity for individual growth and recovery into one of learned helplessness and yet another form of institutionalization, however more humane and pleasant it may be than the stereotypical psychiatric hospital/asylum.

The nature of the therapeutic farm enterprise also creates an unavoidable conflict of interest between the need to maintain a minimal level of consumer use to sustain the enterprise economically and the ethical obligation to help individuals through the recovery process as quickly as they are able to do so. Such conflicts of interest exist across multiple spheres of human enterprise, e.g., increased medical procedures may ensure increased profitability, built-in obsolescence of cars increases profits as replacement becomes necessary. It is not the existence of the conflict of interest that is problematic but rather a lack of awareness or failure to acknowledge it and an unwillingness to implement measures to monitor its effects.

Staffing Considerations

The ability of a therapeutic farm community to provide adequate services to its residents is directly dependent on its ability to recruit and retain qualified staff members. However, employment at a therapeutic farm community differs significantly from employment in a typical mental health services venue. Because the farms are generally situated some distance from a major city, employees will either be required to commute relatively long distances if they wish to live in a large city or live in smaller nearby communities. Additionally, some employees may be required to live on the farm premises if the farm is to maintain 24 h staffing, e.g. in the residential therapeutic/care farm setting. Both small-town residence and residence on the farm suggest that the employee will experience a loss of some degree of privacy that he or she might otherwise have in a larger city or in a day mental health service program, such as a mental health clinic. The features of such employment may narrow the potential pool of interested, qualified applicants and may also increase the potential for staff burnout and/or earlier departure.

Additionally, unlike employment in a mental health clinic where professional staff may interact with their clients for several hours per week at best, employment at a therapeutic farm community requires constant employee-resident interaction throughout the day. As a result of this increased frequency and intensity of contact, staff members and residents are likely to develop a greater familiarity with each other than they might otherwise. Despite this increased familiarity, there continues to be a differential in power between staff and residents, and professional boundaries must be maintained (e.g., CooperRiis Healing Community, 2014, p. 9). As an example, some of the farms have policies indicating that it is inappropriate for staff members to smoke or consume alcohol with residents, to appear intoxicated in front of residents, or to engage in sexual relations with clients (CooperRiis Healing Community, 2014, pp. 9–10). Assurance that appropriate boundaries will be maintained requires continuous efforts to train staff and to foster self-awareness among staff members. Staff members may argue that they are better able to develop a bond with especially isolated clients if they are able to engage with them casually as they smoke or drink together. Nevertheless, staff members essentially serve as models for the clients; one must ask about the wisdom of having a staff member model an unhealthy behavior.

Leadership Challenges

The leadership of a number of the existing therapeutic farms in both the United States and Europe is generally close to the age of retirement, generally about 65–70 years of age. Despite their aging, there has been little succession planning. It appears that some of the existing therapeutic farm communities are unable to identify an already-trained, competent successor should the existing director retire, resign, become ill, or die. This is equally true of many of the members of the farms' governing boards. The development of a line of succession for both administrators and board members and the provision of training for those identified as potential successors are critical to the farms' continuing viability.

Financial Challenges

Both care farms in Europe and therapeutic farm communities in the United States are facing significant financial hurdles. The recession that began during the first decade of the twenty-first century led to significant decreases, often continuing into the present, in government expenditures for mental health services in the United States and in various European countries (Anon, 2012; College of Psychiatrists of Ireland, 2013; Fisher, 2013; National Alliance on Mental Illness, 2011). A withdrawal of governmental financial support could, in some situations, threaten a farm's ability to provide services (e.g., Hough, 2011, p. 11).

In the United States, new federal regulations will increase difficulties associated with obtaining funding for residents' care through the Medicaid health insurance program. The regulations are intended to ensure that individuals receiving long-term services and supports through home and community-based Medicaid-funded programs are part of the community in which they reside and that they are able to access the same conveniences as individuals who are not participating in such Medicaid-funded programs. Under these regulations, which will be fully effective March 17, 2019, funding for care will be available only if the facility is classifiable as a home or community-based setting. This classification requires that the facility meet the following criteria:

- The setting is integrated in and supports full access to the greater community.
- It is selected by the individual from among setting options.
- It ensures individual rights of privacy, dignity and respect, and freedom from coercion and restraint.
- It optimizes autonomy and independence in making life choices.
- It facilitates choice regarding services and who provides them (Centers for Medicare & Medicaid Services, 2014).

Provider-controlled facilities, whether privately or publicly owned, must also meet the following additional requirements to be considered a home or community-based setting:

- The individual has a lease or other legally enforceable agreements providing similar protections.
- The individual has privacy in their unit including lockable doors, choice of room-mates, and freedom to furnish or decorate the unit.
- The individual controls his/her own schedule including access to food at any time.
- The individual can have visitors at any time.
- The setting is physically accessible (Centers for Medicare & Medicaid Services, 2014).

A setting is presumed not to be a home or community-based setting if it is a privately or publicly owned facility that provides inpatient care, is situated on the grounds of or next to a public institution, or separates people receiving Medicaid home and community-based services from those who are not receiving such services. Further, settings that isolate people from receiving home and community-based services from the rest of the community will be deemed ineligible for funding. Under the regulations, settings such as disability-specific farm communities, i.e., therapeutic farm communities, are presumed to isolate people because they are designed specifically for people with a specified type of disability, and on-site staff provides them with services (Centers for Medicare & Medicaid Services, 2014).

A number of these criteria are quite vague, making it difficult to understand how they will be applied. For example, the phrase "the ability to control their schedules and activities, including access to food at any time" suggests that individuals should be able to be served a complete meal at any hour of the day, to have visitors at any time of the day, and to see a therapist or go horseback riding at any hour of the day. Realistically, however, these options would not exist in many communities for anyone at all hours of the day. Does this mean that residents must have the same options available to them as other individuals within the same community? Or does it suggest that service recipients be provided with services beyond what would usually be available? Does the ability of residents to choose their living companions mean that they may choose an individual to live with them who would not be eligible for the farm's services?

Therapeutic farms would likely face great difficulty in their efforts to meet all of these requirements. The structure provided to residents by the therapeutic farms is one of the strengths of the programming. Visits by friends or family members at any time of the day or night, and access to food at any time, may disrupt the structure and the quiet that many residents need. The requirement that a resident be able to lock his or her door at any time could potentially lead to increased hospitalizations in order to proactively prevent the suicide of clients who are experiencing suicidal ideation.

Not surprisingly, these regulations have enormous implications for both individuals seeking services at a therapeutic farm and for the continued viability of the farms. Individuals reliant on Medicaid health insurance who wish to access the services of a therapeutic farm will be unable to do so because of the farm's classification as a non-home or community-based facility. They will, as a result, be forced to rely on either outpatient mental health programs or, if they are sufficiently ill, on inpatient hospital care. Neither avenue holds the prospect of longer-term stable care or the range of services potentially available to them on a therapeutic farm, such as vocational training. Farms are faced with a choice between undergoing significant programmatic restructuring in order to qualify for possible Medicaid funding or foregoing the receipt of Medicaid funding in order to maintain their programming intact. The former approach moves the farm further from its original intent and the underlying premises of moral treatment, while the latter likely limits access to the farm to individuals with sufficient personal, family, or other assets or resources to bear the entire cost of care.

Therapeutic Farms, Politics, and the Public

A lack of understanding of mental illness may lead to ill-advised political decisions that reverberate in the legislative and financial domains. Some politicians may erroneously believe that all care is best delivered on an outpatient basis and/or that "one size fits all" in treating mental illness, whether that one size is medication, hospitalization, isolation from the community, a specific diet or dietary supplement, or genetic engineering. This perspective may stem from a lack of understanding of the wide spectrum of mental illness or may rest entirely on economic considerations (Kliff, 2012).

The stigmatization of both mental illness and those with diagnoses of mental illness may also underlie current political and financial challenges. Just as the poor are often blamed for having created their own poverty and their seeming inability to move beyond it (Ryan, 1971), so too are mentally ill individuals stigmatized and ostracized due to generalized characterizations and unwarranted assumptions (Owen, 2012; Szalavitz, 2012; Tartakovsky, 2013). Media reports highlighting violent crimes committed by individuals suffering from schizophrenia may fail to note the very low rates of violence perpetrated by those with mental illness (Ferriman, 2000; Herrera, 2014). It is not unlikely that such sensationalism leads not only to the exacerbation of public fear and hostility and a reluctance by mentally ill individuals and their families to seek the help they need but also to legislators' unwillingness to fund all but the most restrictive treatment alternatives that are legally permissible.

Virgil Stucker, the current director of CooperRiis, has suggested revisioning residence at a therapeutic farm community as attendance at a "recovery college" (Stucker, 2015). Stucker explained the concept as follows:

> Sometimes, when we want to learn how best to deal with life's challenges, we go away to college. Anthropologists have helped us to understand and value 'liminality' as that place and space that is 'betwixt and between'. The liminal-college experience, for example, helps us to pass from adolescence to adulthood with both social and career skills.
>
> [R]esidential recovery colleges … might provide a means of passage for individuals from their state of being overwhelmed to a state of being able to achieve and sustain their highest level of functioning and fulfillment. One would move through these 'recovery colleges' which would include a system of care that also provides support from community re-integration.

Alternatively, a stay at a therapeutic farm community can be analogized to a sabbatical that provides an academic with a period of respite from the usual administrative and/or teaching responsibilities and offers the assurance of reengagement upon his or her return. Whether likened to either the liminal experience of college or a sabbatical, the reframing might:

1. Normalize the experience as an opportunity to recenter, rebalance, and reenergize
2. Reduce the stigma associated with a request for mental health assistance and physical and self-removal from one's circle of family, friends, and colleagues
3. Emphasize to the larger public that recovery from mental illness is possible
4. Underscore to politicians and legislators the need for and value of alternative approaches to mental health care

Looking Forward

Despite what might seem to be an inexhaustible litany of challenges that singly or multiply exist for many therapeutic farms, numerous strategies exist that can be utilized to transform these challenges into opportunities to grow and move forward as therapeutic communities. These include the reexamination and possible modification of the farm-client relationship, the development of an internal system to monitor conflict of interest, the increased focus on staff training and support, the development of new/increased research capacity and, in the United States in particular, the formation of a therapeutic farm interest group/collaboration/consortium.

The Client-Community Relationship

As discussed in Chap. 3, therapeutic/care farms operate based on a variety of organizational and financial models, some of which more heavily emphasize the entrepreneurial aspects and others the service aspects. As noted earlier, some farms focus primarily on the provision of both mental health services and community support, while others are much more concerned with the provision of housing and social support. Still others appear to be principally concerned with the augmentation of agricultural income, with significantly less attention being paid to the needs of the mentally ill clients. In each of these scenarios, the majority of farms emphasize work and the workday structure as integral components of the therapeutic process but do not provide compensation to their residents/tenants/day workers. This suggests a need to develop on-the-job training programs to avoid the potential exploitation of the residents/clients. Such programs can be used as a vehicle to help residents/clients acquire or relearn both the skills of daily living, such as how to budget and manage personal funds and how to prepare for interviews, and employability skills, such as how to dress for work and how to plan time to arrive at work as expected.

The training provided to therapeutic/care farm staff varies widely across farms depending on the farm's organizational structure, its primary mission, and regulations under which it exists. Training relating to behavioral and mental health may range from none at all to postdoctoral levels. To some degree, these variations are both expected and appropriate. A staff member who is responsible for supervising animal care would not have and should not be expected to have the same type of training as a clinical psychologist conducting group therapy. However, it is potentially dangerous for residents, staff, and the survival of the farm as a therapeutic or care farm if no one has basic training to recognize and provide basic guidance or intervention in physical or mental health emergencies. This may be especially problematic in the case of residential farms that do not maintain 24 h staffing. It is recommended here that all farms develop protocols to address mental and physical health emergencies and that all staff, regardless of the organizational model, be trained to rely on and follow the emergency protocol.

Staffing Considerations

It appears from this author's discussions with staff at various therapeutic farms in the United States and Europe that staff, and particularly those that live on-site in residential programs, encounter limitations on their privacy and questions regarding the appropriateness of staff-client boundaries. These issues exist in significantly greater magnitude than in the usual office-based clinical encounters that are bounded both by time and location. Despite the increased stress that accompanies a relative lack of privacy and continual examination and reexamination of boundaries, few therapeutic or care farms have developed and implemented strategies to proactively assist staff members in managing these complex interactions. Even those that have weekly staff and staff-client meetings often focus their attention on issues related to the individual residents/clients or the general farm community.

The appropriateness of integrating staff self-concerns into community meetings is highly questionable. A variety of strategies can be implemented with staff to provide a suitable, safe venue for the expression of staff concerns and to foster staff self-care. These include regular debriefing sessions and de-stressing activities and opportunities, such as time-out during the day to engage in physical fitness activities, meditation, or other self-care activities.

Financial and Political Issues

Many therapeutic/care farms in Europe are connected through formal and informal consortia that have as their primary focus agriculture-related and/or social service-related concerns. These consortia have frequently developed working relationships with a variety of governmental agencies and, through these interactions, are able to have input into the development of relevant policies and procedures and potentially increase their access to needed resources and funding. In the United States, therapeutic farms for adults with serious mental illness have not established such a consortium. As a result, they have had limited impact on relevant legislation or funding appropriations and allocations. Additionally, it is likely that each farm must spend a growing portion of its budget on marketing because all marketing efforts are conducted singularly. The formation of a consortium may provide a mechanism for the farms to (1) have an increased voice in the development of legislation and regulations related to mental health care and (2) develop and implement a coherent research agenda to examine the effectiveness of the therapeutic farm model as an alternate approach to the treatment of mental illness.

Unanswered Questions: A Suggested Research Agenda

The growing emphasis in recent years by health insurers, health-care practitioners, and consumers of mental health-care services on the use of evidence-based practice suggests that it will become increasingly important for therapeutic/care farms to demonstrate that the services they provide merit the monies charged to receive them if they are to remain sustainable as farms and available to persons who can benefit from such a recovery setting. In psychology, evidence-based practice refers to "the integration of the best available research with clinical expertise in the context of patient characteristics, culture, and preferences" (APA Presidential Task Force on Evidence-Based Practice, 2006, p. 273). The definition closely resembles the definition of evidence-based practice in the field of medicine: "the integration of best research evidence with clinical expertise and patient values" (Institute of Medicine, 2001, p. 147). Accordingly, it is critical that the farms engage in rigorously designed research to develop an evidence base for the model and to identify program components that appear to offer greater effectiveness for specific groups of clients, e.g., by age, presenting diagnosis, or other criteria.

It is critical that we understand what "recovery" means to clients, families, and communities across cultures and societies in order to evaluate the outcomes associated with the therapeutic farm model and its various components. Often, recovery means—to clients, therapeutic farm personnel, families, and the public alike—that an individual is able to live independently and engage in educational endeavors or some form of employment. This approach, however, fails to address various critical issues that confront an individual who is living with mental illness.

The Substance Abuse and Mental Health Services Administration (2012, p. 3) has offered the following definition of recovery: "A process of change through which individuals improve their health and wellness, live a self-directed life, and strive to reach their full potential." Recovery is said to manifest in four domains:

1. Health, meaning the management of one's illness and making informed choices that promote good mental and physical well-being
2. Home, referring to maintaining a stable residence
3. Purpose, signifying the individual's engagement in meaningful daily activities and "the independence, income and resources to participate in society"
4. Community, i.e., maintaining social networks and relationships that foster hope, love, and friendship and furnish support (Substance Abuse and Mental Health Services Administration, 2012, p. 3)

Fulfillment of these four domains is said to be guided by ten principles:

- Hope
- Person driven: self-determination, and self-direction
- Many pathways, meaning that each individual is unique and the recovery process of each is also unique
- Holistic, i.e., encompassing mind, body, spirit, and community and necessitating self-care, transportation, medication, faith, social networks, and support services necessary to facilitate success in these aspects of living
- Peer support
- Relational, recognizing that relationships with family, friends, and/or community may be critical to recovery
- Culture, referring to the need to provide culturally attuned services to aid an individual along the pathway to recovery
- Addresses trauma, recognizing that mental illness is often instigated as the result of unaddressed and unresolved traumatic experience(s)

- Strength/responsibility, suggesting that individuals have a responsibility to care for and promote their own health and communities bear an obligation to reduce the stigma associated with mental illness and promote inclusion
- Respect, emphasizing the need for communities to combat stigma and discrimination and for individuals affected by mental illness to develop a positive and meaningful sense of identity (Substance Abuse and Mental Health Services Administration, 2012, pp. 4–7)

Although this definition may serve as a starting point for research related to recovery, it is not sufficient. The meaning of recovery may vary across and within cultures; the intersectionality of ethnicity, sex, sexual orientation, and/or religion may affect individuals' understanding and process of recovery and healing. An intervention to facilitate recovery will only be embraced by and effective for an individual if it responds to the individual's conceptualization and understanding of recovery, just as individuals will adhere to treatment regimens to the extent that they are congruent with their understandings and representations of their illness (cf. Weisser, Morrow, & Jamer, 2011). As an example, Loue (2011) found in her study involving Puerto Rican women with diagnoses of schizophrenia, bipolar disorder, and major depression that the women variously referred to their symptoms as *nervios* or to their illness as an *ataque de nervios* or "being crazy" (*loca*). Each of these labels carried a different significance, indicating greater or lesser severity of symptoms, greater or lesser insight into the cause of the symptoms and their management, and greater or lesser hope for alleviation and improvement of one's condition.

References

Anon. (2012). The impact of the economic downturn on public mental health systems. *Psychiatric Times*, Feb 8. http://www.psychiatrictimes.com/articles/impact-economic-downturn-public-mental-health-systems. Accessed 31 July 2015.

APA Presidential Task Force on Evidence-Based Practice. (2006). Evidence-based practice in psychology. *American Psychologist, 61*(4), 271–285.

Centers for Medicare & Medicaid Services. (2014, January 10). *Fact sheet: Summary of key provisions of the home and community-based services (HCBS) settings final rule (CMS 2249-F/2296-F)*. Baltimore, MD: U.S. Department of Health & Human Services. http://www.medicaid.gov/medicaid-chip-program-information/by-topics/long-term-services-and-supports/home-and-community-based-services/downloads/hcbs-setting-fact-sheet.pdf. Accessed 10 May 2015.

College of Psychiatrists of Ireland. (2013). *Pre-budget 2014 submission re mental health services*, September. http://www.irishpsychiatry.ie/Libraries/External_Events_Documents/CPsychI_Pre_Budget_Submission_FINAL_2_4.sflb.ashx/. Accessed 31 July 2015.

CooperRiis Healing Community. (2014). *CooperRiis employee handbook*. Mill Spring, NC: Author.

Ferriman, A. (2000). The stigma of schizophrenia. *British Medical Journal, 320*(7233), 522.

Fisher, N. (2013). Mental health loses funding as government continues shutdown. *Forbes*, Oct 10. http://www.forbes.com/sites/theapothecary/2013/10/10/mental-health-loses-funding-as--government-continues-shutdown/. Accessed 31 July 2105.

Frese, F. J., III, Stanley, J., Kress, K., & Vogel-Scibilia, S. (2001). Integrating evidence-based practices and the recovery model. *Psychiatric Services, 52*(11), 1462–1468.

Herrera, V. (2014). Double murder drawing attention to schizophrenia. *NBC San Diego*, Dec 1. http://www.nbcsandiego.com/news/local/Double-Murder-Drawing-Attention-to--Schizophrenia-284273671.html. Accessed 31 July 2015.

Hough, J. (2011). Mental health project under threat over funding loss. *Irish Examiner*, Mar 11.

Institute of Medicine. (2001). *Crossing the quality chasm: A new health system for the 21st century*. Washington, DC: National Academies Press.

Kliff, S. (2012). Seven facts about America's mental health-care system. *The Washington Post*, Dec 17. http://www.washingtonpost.com/blogs/wonkblog/wp/2012/12/17/seven-facts-about-americas-mental-health-care-system/. Accessed 31 July 2015.

Loue, S. (2011). *"My nerves are bad" ("Mis nervios estan malos"): Puerto Rican women managing mental illness and HIV risk*. Nashville, TN: Vanderbilt University Press.

Munetz, M. R., & Frese, F. J. (2001). Getting ready for recovery: Reconciling mandatory treatment with the recovery vision. *Psychiatric Rehabilitation Journal, 25*(1), 35–42.

National Alliance on Mental Illness. (2011). *State mental health cuts: A national crisis. A report by the National Alliance on Mental Illness*, March. https://www2.nami.org/ContentManagement/ContentDisplay.cfm?ContentFileID=125018. Accessed 31 July 2015.

Owen, P. R. (2012). Portrayals of schizophrenia by entertainment media: A content analysis of contemporary movies. *Psychiatric Services, 63*(7), 655–659.

Ryan, W. (1971). *Blaming the victim*. New York: Random House, Inc.

Stucker, V. (2015). *Recovery colleges, not asylums: Looking into the past for solutions*. http://www.peteearley.com/2015/05/11/recovery-colleges-not-asylums-looking-into-the-past-for-solutions/. Accessed 26 July 2015.

Substance Abuse and Mental Health Services Administration. (2012). *SAMHSA's working definition of recovery: 10 guiding principles of recovery*. http://store.samhsa.gov/shin/content/PEP12-RECDEF/PEP12-RECDEF.pdf. Accessed 8 August 2015.

Szalavitz, M. (2012). After Aurora, questions about mass murder and mental illness. *Time*, July 31. http://healthland.time.com/2012/07/31/mass-murder-and-mental-illness-the-interplay-of-stigma-culture-and-disease/. Accessed 31 July 2015.

Tartakovsky, M. (2013). Media's damaging depictions of mental illness. *PsychCentral*. http://psychcentral.com/lib/medias-damaging-depictions-of-mental-illness/?all=1. Accessed 31 July 2015.

Weisser, J., Morrow, M., & Jamer, B. (2011). *A critical exploration of social inequities in the mental health recovery literature*. Vancouver, BC: Centre for the Study of Gender, Social Inequities and Mental Health (CGSM). http://www.socialinequities.ca. Accessed 8 August 2015.

Index

© Springer International Publishing Switzerland 2016
S. Loue, *Therapeutic Farms*, SpringerBriefs in Social Work,
DOI 10.1007/978-3-319-13539-7

The manufacturer's authorised representative in the EU is Springer
Nature Customer Service Centre GmbH, Europaplatz 3, 69115 Heidelberg,
Germany. If you have any concerns regarding our products, please
contact ProductSafety@springernature.com

Printed and bound by CPI Group (UK) Ltd, Croydon, CR0 4YY

27/04/2026

02097573-0014